INTRODUCTION TO WATER ENGINEERING, HYDROLOGY, AND IRRIGATION

INTRODUCTION TO WATER ENGINEERING, HYDROLOGY, AND IRRIGATION

Mohammad Albaji

CRC Press
Taylor & Francis Group
Boca Raton London New York

CRC Press is an imprint of the
Taylor & Francis Group, an **informa** business

First edition published 2022
by CRC Press
6000 Broken Sound Parkway NW, Suite 300, Boca Raton, FL 33487-2742

and by CRC Press
4 Park Square, Milton Park, Abingdon, Oxon, OX14 4RN

CRC Press is an imprint of Taylor & Francis Group, LLC

© 2022 Taylor & Francis Group, LLC

Library of Congress Cataloging-in-Publication Data
Names: Albaji, Mohammad, author.
Title: Introduction to water engineering, hydrology, and irrigation / Mohammad Albaji.
Description: First edition. I Boca Raton : CRC Press, 2022. I Includes bibliographical references and index. I
Summary: "This book is designed as an undergraduate text for water and environmental engineering courses and as preliminary reading for postgraduate courses in water and environmental engineering- including introductory coverage of irrigation and drainage, water resources, hydrology, hydraulic structures, and more. The text and exercises have been classroom tested by undergraduate water and environmental engineering students and are augmented by material prepared for extramural short courses. It covers basic concepts of agricultural irrigation and drainage, including planning and design, surface intakes, economics, environmental impacts wetlands, and legal issues.
Features: Numerous illustrations throughout to clarify the concepts presented Examines and compares the advantages and disadvantages of several methods of irrigation practice Explains the integral components including pumps, filters, piping, valves, and more Considers fertilizer application and nutrient management This comprehensive and well-illustrated book will be of great interest to students, professionals, and researchers involved with all aspects of water engineering, hydrology, and irrigation"-- Provided by publisher.
Identifiers: LCCN 2021059518 (print) I LCCN 2021059519 (ebook) I ISBN 9781032275925 (hardback) I ISBN 9781032276083 (paperback) I ISBN 9781003293507 (ebook)
Subjects: LCSH: Hydraulic engineering--Textbooks. I Hydrology--Textbooks. I Irrigation--Textbooks. I Irrigation engineering--Textbooks.
Classification: LCC TC145 .A52 2022 (print) I LCC TC145 (ebook) I DDC 627--dc23/eng/20220217
LC record available at https://lccn.loc.gov/2021059518
LC ebook record available at https://lccn.loc.gov/2021059519

ISBN: 978-1-032-27592-5 (hbk)
ISBN: 978-1-032-27608-3 (pbk)
ISBN: 978-1-003-29350-7 (ebk)

DOI: 10.1201/9781003293507

Typeset in Times
by MPS Limited, Dehradun

Contents

Contents

Preface

Water engineering, hydrology, and irrigation comprise a wide range of different majors that are developing and changing rapidly, so students and specialists of these majors need to use the most up-to-date texts and papers. This can be achieved through reading the original texts. This book is designed as a text for undergraduate water engineering, hydrology, and irrigation courses and as preliminary reading for postgraduate courses in water engineering, hydrology, and irrigation. It is hoped that it will also be of value to specialists, experts, and engineers already in the field and to students preparing for the M.Sc. and PhD examinations. The texts and exercises are based on my lecture courses to undergraduate water and environmental engineers augmented by material prepared for extramural short courses. Wherever possible, illustrations have been used to clarify the texts.

The book is divided into 16 chapters and is intended for students, researchers, and professionals working on various aspects of water engineering, hydrology, and irrigation.

An honest effort has been made to check the scientific validity and justification of each chapter through several iterations. We, the authors and the publisher, have put together a comprehensive reference for students of water, hydrology, and irrigation engineering with a belief that this book will be of immense use to present and future colleagues who teach, study, and/or practice in this particular field. We would welcome comments from anyone that will help us improve any future editions.

Mohammad Albaji, PhD
Associate Professor
Department of Irrigation and Drainage
Faculty of Water & Environmental Engineering
Shahid Chamran University of Ahvaz
Ahvaz, Iran

Author

Mohammad Albaji is a faculty member at Shahid Chamran University of Ahvaz, Iran. He earned his PhD in Irrigation and Drainage Engineering from Shahid Chamran University of Ahvaz, Iran. Dr. Albaji is active in the fields of irrigation and drainage, precision irrigation, agricultural water management, water productivity, water and soil salinity and alkalinity, soil science, land suitability for irrigation, land suitability for crops, and has more than 100 publications in reputed journals, books, and refereed conferences. He has published three Handbooks; *"Handbook of Irrigation System Selection for Semi-Arid Regions"* in 2020, *"Handbook of Technical Terms of Soil and Water Engineering"* in 2021, and *"Handbook of Irrigation Hydrology and Management"* in 2022. He has been working as a reviewer for many reputed journals such *as Agricultural Water Management, Computers and Electronics in Agriculture, Water Resources Management, Environmental Earth Sciences, Environment, Development and Sustainability, Transaction of the Royal Society of South Africa, Clean Soil, Air, Water,* etc. He had several executive posts such as dean of Jundishapur's Water and Energy Research Institute (Shahid Chamran University of Ahvaz; 2016–2018), head of Soil Science Department (Khuzestan Water and Power Authority; 2000–2012), etc. He has membership in the International Center for Biosaline Agriculture (ICBA), American Society of Civil Engineering (ASCE), etc.

Acknowledgments

I am grateful to the Research Council of Shahid Chamran University of Ahvaz for financial support (GN: SCU.WI1400.280).

1 Hydraulic engineering

READING FOR COMPREHENSION

Hydraulic engineering as a sub-discipline of civil engineering is concerned with the flow and conveyance of fluids, principally water and sewage. One feature of these systems is the extensive use of gravity as the motive force to cause the movement of the fluids. This area of civil engineering is intimately related to the design of bridges, dams, channels, canals, and levees, and to both sanitary and environmental engineering.

Hydraulic engineering is the application of fluid mechanics principles to problems dealing with the collection, storage, control, transport, regulation, measurement, and use of water. Before beginning a hydraulic engineering project, one must figure out how much water is involved. A hydraulic engineer is concerned with the transport of sediment by the river, the interaction of the water with its alluvial boundary, and the occurrence of scour and deposition. "The hydraulic engineer actually develops conceptual designs for the various features which interact with water such as spillways and outlet works for dams, culverts for highways, canals and related structures for irrigation projects, and cooling-water facilities for thermal power plants" (Figures 1.1–1.3).

FUNDAMENTAL PRINCIPLES

A few examples of the fundamental principles of hydraulic engineering include fluid mechanics, fluid flow, behavior of real fluids, hydrology, pipelines, open channel hydraulics, mechanics of sediment transport, physical modeling, hydraulic machines, and drainage hydraulics.

FLUID MECHANICS

Fundamentals of hydraulic engineering define hydrostatics as the study of fluids at rest. In a fluid at rest, there exists a force, known as pressure, that acts upon the fluid's surroundings. This pressure, measured in N/m^2, is not constant throughout the body of fluid. Pressure, p, in a given body of fluid, increases with an increase in depth. Where the upward force on a body acts on the base and can be found by equation:

$$p = \rho g y$$

where
 ρ = density of water
 g = specific gravity
 y = depth of the body of liquid

DOI: 10.1201/9781003293507-1

FIGURE 1.1 Hydraulic flood retention basin (HFRB); (https://en.wikipedia.org/wiki/File:Hydraulic_Flood_Retention_Basin.jpg).

Rearranging this equation gives you the pressure head $p/\rho g = y$. Five basic devices for pressure measurement are a piezometer, manometer, differential manometer, Bourdon gauge, as well as an inclined manometer.

As Prasuhn states:

On undisturbed submerged bodies, pressure acts along all surfaces of a body in a liquid, causing equal perpendicular forces in the body to act against the pressure of the liquid. This reaction is known as equilibrium. More advanced applications of pressure are those on plane surfaces, curved surfaces, dams, and quadrant gates, just to name a few.

BEHAVIOR OF REAL FLUIDS

Real and Ideal Fluids

The main difference between an ideal fluid and a real fluid is that for ideal flow $p_1 = p_2$ and for real flow $p_1 > p_2$. Ideal fluid is incompressible and has no viscosity. Real fluid has viscosity. Ideal fluid is only an imaginary fluid as all fluids that exist have some viscosity.

FIGURE 1.2 View from Ahvaz cable bridge, Ahvaz, Iran (By Mohammad Albaji).

FIGURE 1.3 Riprap lining a lakeshore; (https://en.wikipedia.org/wiki/Riprap).

Viscous Flow

A viscous fluid will deform continuously under a shear force, whereas an ideal fluid doesn't deform.

Laminar Flow and Turbulence

The various effects of disturbance on a viscous flow are stable, transition, and unstable.

Bernoulli's Equation

For an ideal fluid, Bernoulli's equation holds along streamlines.

$$p/\rho g + u^2/2g = p_1/\rho g + u_1^2/2g = p_2/\rho g + u_2^2/2g$$

Boundary Layer

Assuming a flow is bounded on one side only, and that a rectilinear flow passing over a stationary flat plate that lies parallel to the flow, the flow just upstream of the plate has a uniform velocity. As the flow comes into contact with the plate, the layer of fluid actually "adheres" to a solid surface. There is then a considerable shearing action between the layer of fluid on the plate surface and the second layer of fluid. The second layer is therefore forced to decelerate (though it is not quite brought to rest), creating a shearing action with the third layer of fluid, and so on. As the fluid passes further along the plate, the zone in which shearing action occurs tends to spread further outwards. This zone is known as the "boundary layer." The flow outside the boundary layer is free of shear and viscous-related forces so it is assumed to act like an ideal fluid. The intermolecular cohesive forces in a fluid are not great enough to hold fluid together. Hence a fluid will flow under the action of the slightest stress and flow will continue as long as the stress is present. The flow inside the layer can be either viscous or turbulent, depending on the Reynolds number.

Applications

Common topics of design for hydraulic engineers include hydraulic structures such as dams, levees, water distribution networks, water collection networks, sewage collection networks, stormwater management, sediment transport, and various other topics related to transportation engineering and geotechnical engineering. Equations developed from the principles of fluid dynamics and fluid mechanics are widely utilized by other engineering disciplines such as mechanical, aeronautical, and even traffic engineers.

Related branches include hydrology while related applications include hydraulic modeling, flood mapping, catchment flood management plans, shoreline management plans, estuarine strategies, coastal protection, and flood alleviation.

Hydraulic Structure

A hydraulic structure is a structure submerged or partially submerged in any body of water, which disrupts the natural flow of water. They can be used to divert, disrupt or completely stop the flow. An example of a hydraulic structure would be a dam, which slows the normal flow rate of the river in order to power turbines. A hydraulic structure can be built in rivers, a sea, or any body of water where there is a need for a change in the natural flow of water.

Hydraulic structures may also be used to measure the flow of water. When used to measure the flow of water, hydraulic structures are defined as a class of specially shaped, static devices over or through which water is directed in such a way that under free-flow conditions at a specified location (point of measurement) a known level to flow relationship exists. **Hydraulic structures** of this type can generally be divided into two categories: flumes and weirs.

WATER DISTRIBUTION NETWORK

The product, delivered to the point of consumption, is called potable water if it meets the water quality standards required for human consumption.

The water in the supply network is maintained at positive pressure to ensure that water reaches all parts of the network, that a sufficient flow is available at every take-off point and to ensure that untreated water in the ground cannot enter the network. The water is typically pressurized by pumps that pump water into storage tanks constructed at the highest local point in the network. One network may have several such service reservoirs (Figure 1.4).

FIGURE 1.4 Ghadir water conveyance and distribution network, Ahvaz, Iran (By Mohammad Albaji).

In small domestic systems, the water may be pressurized by a pressure vessel or even by an underground cistern (the latter however does need additional pressurizing). This eliminates the need for a water tower or any other heightened water reserve to supply the water pressure.

These systems are usually owned and maintained by local governments, such as cities or other public entities, but are occasionally operated by a commercial enterprise (see water privatization). Water supply networks are part of the master planning of communities, counties, and municipalities. Their planning and design require the expertise of city planners and civil engineers, who must consider many factors such as location, current demand, future growth, leakage, pressure, pipe size, pressure loss, firefighting flows, etc. using pipe network analysis and other tools.

As water passes through the distribution system, the water quality can degrade by chemical reactions and biological processes. Corrosion of metal pipe materials in the distribution system can cause the release of metals into the water with undesirable aesthetic and health effects. Release of iron from unlined iron pipes can result in customer reports of "red water" at the tap. Release of copper from copper pipes can result in customer reports of "blue water" and/or a metallic taste. Release of lead can occur from the solder used to join copper pipe together or from brass fixtures. Copper and lead levels at the consumer's tap are regulated to protect consumer health.

Utilities will often adjust the chemistry of the water before distribution to minimize its corrosiveness. The simplest adjustment involves control of pH and alkalinity to produce water that tends to passivity corrosion by depositing a layer of calcium carbonate. Corrosion inhibitors are often added to reduce the release of metals into the water. Common corrosion inhibitors added to the water are phosphates and silicates.

The maintenance of biologically safe drinking water is another goal in water distribution. Typically, a chlorine-based disinfectant such as sodium hypochlorite or monochloramine is added to the water as it leaves the treatment plant. Booster stations can be placed within the distribution system to ensure that all areas of the distribution system have adequate sustained levels of disinfection.

SEWAGE COLLECTION NETWORKS

Sewage collection systems transport sewage through cities and other inhabited areas to sewage treatment plants to protect public health and prevent disease. Sewage is treated to control water pollution before discharge to surface waters.

STORMWATER MANAGEMENT

Stormwater is rainwater and melted snow that runs off streets, lawns, and other sites. When stormwater is absorbed into the ground, it is filtered and ultimately replenishes aquifers or flows into streams and rivers. In developed areas, however, impervious surfaces such as pavement and roofs prevent precipitation from naturally soaking into the ground. Instead, the water runs rapidly into storm drains, sewer systems, and drainage ditches and can cause: downstream flooding, stream bank erosion, increased turbidity (muddiness created by stirred up sediment) from erosion, habitat destruction, changes in the streamflow hydrograph (a graph that displays the flow rate of a stream over a period of time), combined sewer overflows, infrastructure damage, contaminated streams, rivers, and coastal water.

SEDIMENT TRANSPORT

Sediment transport is the movement of solid particles (sediment), typically due to a combination of gravity acting on the sediment, and/or the movement of the fluid in which the sediment is entrained. Sediment transport occurs in natural systems where the particles are clastic rocks (sand, gravel, boulders, etc.), mud, or clay; the fluid is air, water, or ice; and the force of gravity acts to move the particles along the sloping surface on which they are resting. Sediment transport due to fluid motion occurs in rivers, oceans, lakes, seas, and other bodies of water due to currents and tides. Transport is also caused by glaciers as they flow, and on terrestrial surfaces under the influence of wind. Sediment transport due only to gravity can occur on sloping surfaces in general, including hillslopes, scarps, cliffs, and the continental shelf – continental slope boundary (Figure 1.5).

FIGURE 1.5   Sediment transported by Karun River, Ahvaz, Iran (By Mohammad Albaji).

EXERCISES

A. **Read each statement and decide whether it is true or false. Write "T" for true and "F" for false statements.**

TF1.   Hydraulic engineering is the application of fluid mechanics principles to problems dealing with water subjects.

TF2.   Hydraulic engineering as a sub-discipline of mechanical engineering is concerned with the flow and conveyance of fluids, principally water and sewage.

TF3.   The hydraulic engineer is concerned with the transport of sediment by the river.

TF4.   The water in the supply network is maintained at negative pressure.

TF5.   When water passes through the distribution system, the water quality can degrade by chemical reactions processes.

B. **Circle a, b, c, or d which best completes the following items.**

1.   One feature of hydraulic engineering is the extensive use of .... as the motive force to cause the movement of the fluids.
   a. gravity
   b. pressure
   c. electricity
   d. slope

2.   In a .... at rest, there exists a force, known as pressure that acts upon the fluid's surroundings.
   a. water
   b. fluid
   c. gas
   d. liquid

3.   On undisturbed submerged bodies, pressure acts along all surfaces of a body in a liquid, causing equal .... forces in the body to act against the pressure of the liquid.
   a. pressure
   b. cohesion
   c. adhesion
   d. perpendicular

4.   An example of a hydraulic structure would be a dam, which slows the .... flow rate of the river in order to power turbines.
   a. slow
   b. normal
   c. speed
   d. a, b, and c

5. .... is rainwater and melted snow that runs off streets, lawns, and other sites.
   a. Wastewater
   b. Surface water
   c. Stormwater
   d. Reuse water

C. **Match the sentence halves in Column I with their appropriate halves in Column II. Insert the letters a, b, c ... in the parentheses provided. There are more sentence halves in Column II than required.**

Column I	Column II
1. Sediment transport is	() **a.** the water may be pressurized by a pressure vessel or even by an underground cistern.
2. Maintenance of a biologically safe drinking water is	() **b.** may also be used to measure the flow of water.
3. In small domestic systems,	() **c.** of the fluid in which the sediment is entrained.
4. A hydraulic structure is	() **d.** the movement of solid particles, typically due to a combination of gravity acting on the sediment.
5. Hydraulic structures	() **e.** another goal in water distribution.
	() **f.** structure submerged or partially submerged in any body of water, which disrupts the natural flow of water.
	() **g.** of the fundamental principles of hydraulic engineering include fluid mechanics, fluid flow.
	() **h.** treated to control water pollution before discharge to surface waters.

D. **Cross out the word or words that make each statement false, and write the word or words that make each statement true in the blank.**

1. The main difference between an ideal fluid and a real fluid is that for real flow $p_1 = p_2$ and for ideal flow $p_1 > p_2$.
   ....

2. Ideal fluid is compressible incompressible and has viscosity.
   ....

3. Water supply networks are not part of the master planning of communities, counties, and municipalities.
   ....

4. Sewage collection systems transport sewage through cities and other inhabited areas to rivers.
   ....

E. **Give answers to the following questions.**
   1. What is hydraulic engineering?
   2. What are the fundamental principles of hydraulic engineering?
   3. What are the applications of hydraulic engineering?
F. **For each word on the left, there are three meanings provided. Put a check mark (√) next to the choice which has the closest meaning to the word given.**

1. Submerged	above of water	below of water	over of water
2. Sewage	untreated water	treated water	drainage water
3. Sediment	salt particles	fine particles	solid particles
4. Fluid	liquid or gas	liquid or solid	solid or gas
5. Turbulent flow	move parallel	move randomly	move vertical

G. **Fill in the blanks with the appropriate words from the following list.**

open channel flow   pressurized   pneumatics   liquids   hydraulic
modules   engineering   free surface   fluid power   fluid mechanics

Hydraulics is a topic in applied science and .... dealing with the mechanical properties of .... or fluids. At a very basic level, hydraulics is the liquid version of .... provides the theoretical foundation for hydraulics, which focuses on the engineering uses of fluid properties. In ...., hydraulics is used for the generation, control, and transmission of power by the use of .... liquids. .... topics range through some parts of science and most of engineering ...., and cover concepts such as pipe flow, dam design, fluidics and fluid control circuitry, pumps, turbines, hydropower, computational fluid dynamics, flow measurement, river channel behavior, and erosion.

Free surface hydraulics is the branch of hydraulics dealing with .... flow such as occurring in rivers, canals, lakes, estuaries, and seas. Its sub-field .... studies the flow in open channels (Figure 1.6).

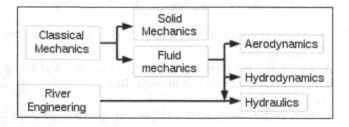

FIGURE 1.6 Hydraulics and other studies.

H. **Read this passage and then circle a, b, c, or d which best completes the following items.**

After students understand the basic principles of hydraulics, some teachers use a hydraulic analogy to help students learn other things. For example:

1. The MONIAC Computer uses water flowing through hydraulic components to help students learn about economics.
2. The thermal-hydraulic analogy uses hydraulic principles to help students learn about thermal circuits.
3. The electronic–hydraulic analogy uses hydraulic principles to help students learn about electronics.

   The main purpose of this passage is ....

a. introduction of hydraulic models.
b. hydraulic principles.
c. learning of hydraulic.
d. hydraulic and computer.

I. **Translate the following passage into your mother language. Write your translation in the space provided.**

The qanat technology is known to have been developed by the Persian people sometime in the early 1st millennium BC, and seen in the fourth millennium BC, and spread from there slowly westward and eastward.

Qanats are constructed as a series of well-like vertical shafts, connected by gently sloping tunnels. Qanats tap into subterranean water in a manner that efficiently delivers large quantities of water to the surface without need for pumping. The water drains by gravity, with the destination lower than the source, which is typically an upland aquifer. Qanats allow water to be transported over long distances in hot dry climates without loss of much of the water to evaporation.

It is very common in the construction of a qanat for the water source to be found below ground at the foot of a range of foothills of mountains, where the water table is closest to the surface. From this point, the slope of the qanat is maintained closer to level than the surface above, until the water finally flows out of the qanat above ground. To reach an aquifer, qanats must often extend for long distances.

Qanats are sometimes split into an underground distribution network of smaller canals called kariz. Like qanats, these smaller canals are below ground to avoid contamination. In some cases, water from a qanat is stored in a reservoir, typically with night flow stored for daytime use. An Ab Anbar is an example of a traditional qanat-fed reservoir for drinking water in Persian antiquity.

The qanat system has the advantage of being resistant to natural disasters such as earthquakes and floods, and to deliberate destruction in

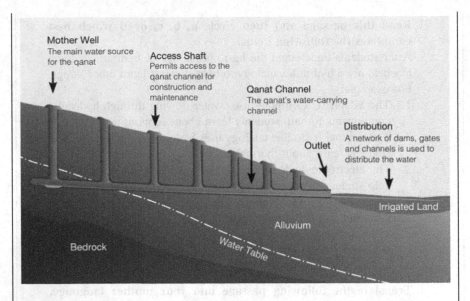

FIGURE 1.7   Cross-section of a qanat; (https://en.wikipedia.org/wiki/Qanat#Iran).

war. Furthermore, it is almost insensitive to the levels of precipitation, delivering a flow with only gradual variations from wet to dry years (Figure 1.7).

J. Copy the technical terms and expressions used in this lesson. Then find your mother language equivalents of those terms and expressions and write them in the spaces provided.

Technical term	Mother language equivalent
....	....
....	....
....	....
....	....
....	....
....	....
....	....
....	....
....	....
....	....

BIBLIOGRAPHY

Cassidy, J., Chaudhry, J., Hanif, M., & Roberson, J. (1998). *Hydraulic Engineering*. John Wiley & Sons, 1998.

Finnemore, E.J. & Franzini, J. (2002). *Fluid Mechanics with Engineering Applications*. McGraw-Hill, 2002.

Goldsmith, E. The Qantas of Iran.

http://www.constructionweekonline.com/article-21843-honeywell-wins-118m-kuwait-water-deal

https://en.wikipedia.org/wiki/File:Hydraulic_Flood_Retention_Basin.jpg

https://en.wikipedia.org/wiki/Hydraulic engineering

https://en.wikipedia.org/wiki/Hydraulics

https://en.wikipedia.org/wiki/Qanat#Iran

https://en.wikipedia.org/wiki/Riprap

https://en.wikipedia.org/wiki/Switzerland

Hydraulic structure: flumes and weirs.

Hydraulic structures United States Army Corps of Engineers.

Hydroelectric Power Plants in Iran. IndustCards. Retrieved 13 January 2015.

Kareez (kariz, karez, qanat). Heritage Institute.

Prasuhn, A.L. (1987). *Fundamentals of Hydraulic Engineering*. New York: Holt, Rinehart, and Winston, 1987.

Qanats (PDF). *Water History*.

The qanats of Iran. *Bart.nl*.

Wilson, A. (2008). John Peter Oleson, ed. Hydraulic Engineering and Water Supply. *Handbook of Engineering and Technology in the Classical World*. New York: Oxford University Press. p. 291f. ISBN 978-0-19-973485-6.

GLOSSARY OF TERMS

Boundary layer: The layer of fluid closest to the surface of a solid past at which the fluid flows: it has a lower rate of flow than the bulk of the fluid because of its adhesion to the solid.

Fluid: A substance, as a liquid or gas, that is capable of flowing and that changes its shape at a steady rate when acted upon by a force tending to change its shape.

Fluid mechanics: The branch of physics that involves the study of fluids (liquids, gases, and plasmas) and the forces on them.

Hydraulic engineering: The branch of civil engineering dealing with the use and control of water in motion.

Hydraulic structure: Hydraulic structure is a structure submerged or partially submerged in any body of water, which disrupts the natural flow of water.

Laminar flow: The flow of a viscous fluid in which particles of the fluid move in parallel layers, each of which has a constant velocity but is in motion relative to its neighboring layers.

Sediment transport: Sediment transport is the movement of solid particles (sediment), typically due to a combination of gravity acting on the sediment, and/or the movement of the fluid in which the sediment is entrained.

Stormwater: Stormwater is rainwater and melted snow that runs off streets, lawns, and other sites.

Turbulent flow: The motion of a fluid having local velocities and pressures that fluctuate randomly.

Viscous flow: A type of fluid flow in which there is a continuous steady motion of the particles, the motion at a fixed point always remaining constant.

2 River engineering

READING FOR COMPREHENSION

River engineering is the process of human planned actions in the course, characteristics, or flow of a river for some certain uses. Before recorded history, humans have intervened in the natural conditions of rivers for some purposes including managing the water resources, protecting against flooding, make passage along or across rivers easier. Rivers have been used as a source of hydropower since Roman times. With the spread of environmental concerns in the late 20th century, today the main purpose of some river engineering projects is protecting environment with some actions like the restoration or protection of natural characteristics and habitats (Figure 2.1).

CHARACTERISTICS OF RIVERS

Size

Tidal limit and average freshwater discharge can affect the rivers size but the main factors that determine the size of a river are the extent of basin and the amount of rain over the basin.

Basin

The basin of a river is part of a country that is confined to a watershed where rain falls to the lowest point. River basins size depended on the form of the country and the extent of the largest river basin in every country is proportionate to the size of its continent, the position of that relative to the hilly regions and the sea and the distance between the headwater and the outlet where the rivers flow into the sea.

Rate of Flow

The flow rate of rivers approximately corresponds to their gradient. When several rivers have the same gradient, the river that has the bigger size has the faster flow because reduce in flow speed caused by fraction is less compared to smaller rivers. The slope of a river depends on the fall of the land it crosses. Rivers in highest elevation of their basins move fast and torrents with a very variable flow. When the stream arrives plains regions in the latter part of its traverse, usually become quite gentle and have comparatively regular discharge.

METHODS

River engineering methods are divided into two categories based on the overall purpose. When the purpose of river engineering is to improve flow, especially in

DOI: 10.1201/9781003293507-2

FIGURE 2.1   River engineering channel; (https://restorerivers.eu/wiki/images/d/d9/River_
Engineering_Channel.jpg).

flood conditions, is called channelization and when the purpose is to hold back the
flow, primarily for navigation purposes, is known as canalization.

Channelization

For increasing the effective fall, the only solution is decreasing the channels length
by replacing winding sections with straight cuts, however it reduce the channels
capacity. Large rivers with high rate of flow have tendency to push the banks back
and form again a sinuous channel, therefore, it is too hard to preserve a straight cut.

Obstructions in the bed of a river depend on their size, impede the flow, raise the
level of the river above it, produce the additional artificial fall necessary to transport
the stream through the restricted channel, and consequently decreases the effective
fall. The elimination of obstructions is an accessible and advantageous way for
increasing the discharge capacity in channel, which results in lower flood-level
upstream. Obstructions may be natural such as trunks of trees, rocks, and gravel
mass or be artificial, as a result of human actions such as wide piers of bridges, solid
weirs, sluice gates for mills, mining refuse, and fish traps.

Stream gauges can be helpful in flood control by installing in a large river and its
tributaries at proper points. Gauges that record parameters include water elevation
at the different stations, the rise of the floods in the various tributaries, the length of
time that they take to transport to certain stations on the main river, and their effect
on the floods height at these places. Using these recorded parameters, the time of
arrival and maximum flood height at any station on the main river can be predicted
with considerable accuracy a few days ago. Sharing this information makes weir-
keepers able to open the movable weirs for flood passage on suitable time and the
people of those areas receive warning of the impending flood (Figure 2.2).

FIGURE 2.2 A channelizes stream in Western Pennsylvania; (https://waterlandlife.org/wildlife-pnhp/species-at-risk-in-pennsylvania/freshwater-mussels/pnhp-species at-risk-mussel-surveys mw 4-channellzed-stream-400x300/).

ADVANTAGES AND DISADVANTAGES

The advantages of channelization are:

- Creation of a suitable space for navigation by larger ships.
- Directing river water to a specific area for agriculture.
- Flood control, giving a stream a sufficiently large and deep channel.
- Declines natural erosion.

The disadvantages of channelization are:

- Loss of wetlands as a filter for much of the surface fresh water and a habitat for many species of wildlife.
- Increase in soil erosion and flooding downstream because channelized streams straightened in most cases that cause more volumes of water moving faster than normal streams.
- Decreased fish population in the river because of decline in habitat, loss of riffles and pools, more swing in stream levels and water temperature, and shifting substrates.

CANALIZATION

Rivers that have a steep slope or their flow reduced at downstream cannot provide adequate depth for navigation just by actions that regulate the flow. The usual level of rivers in summer should be raised by impounding the flow with weirs at intervals across the channel, while a lock has to be provided alongside the weir, or in a side channel to create a suitable course for the vessels cross. A river that converted to replacement for fairly level canal pounds rising in steps upstream makes still-water navigation comparable to a canal. Navigation may be disrupted by floods that reach the top of the locks or in cold climates with severe frosts (Figure 2.3).

REGULATION WORKS (FLOW AND DEPTH CONTROL)

When rivers move toward the sea, they got a significant decrease in their slope and a progressive increase in extent of their drain basin because of consecutive entry of tributaries. Therefore, their flow becomes slowdown but their discharge volume becomes larger and less subject to abrupt variations, as a result, they become more useful for navigation purposes. Finally, large rivers, in normal conditions, often provide natural main ways for navigation in the lower sections of their course. River engineering works for the improvement of navigation condition can be done usefully just in cases that large rivers have a fair slope and adequate discharge at their lowest stage. Large amounts of fall in rivers course is an obstacle for navigation owing to considerable variations in water level. In the dry season when the discharge is quite low, it is impossible to furnish adequate water depth in the low-water channel.

FIGURE 2.3 A navigation lock in Yorkshire, United Kingdom; (https://www.waymarking.com/ waymarks/WMVNVE_River_Don_Navigation_Mexborough_Top_Lock_Mexborough_UK).

To provide a uniform depth in a river by lowering the shoals that are impediments to the channel, the shoal material must be considered. The lowering of a soft shoal by dredging merely create an interim deepening and then it forms again because of process such as deposit from reduce in flow velocity, decrease in slope, channel widening, loss in scour concentration of the stream moving from one concave bank to the next on the opposite side. The elimination of the rocky blocks above the rapids results in depth increasing and flow equalizing but makes lowering of the river at the rapids by facilitating the efflux, which can produce new shoals at the low stage of the river. However, elimination of the narrow rocky reefs or other hard shoals that stretch across the river bottom and obstructions the erosion by the soft materials current that forming the river bed can make river able to increase the depth of its bed by natural scour, which result in permanent improvement (Figure 2.4).

The possibility of furnishing a suitable condition for navigation in a river in dry season with low discharge and deficit scour depends on the maintained depth at the lowest stage of channel. A usual method for solving this problem in the low-water channels is closing subsidiary low-water channels with dikes across them and decreasing the width of channel at the low stage to concentrate all of the flow into a central channel.

ESTUARINE WORKS

In addition, navigation needs a stable and navigable channel that continuous from the navigable river to deep water at the mouth of the estuary. The actions and reactions between river flow and tide should be modeled using computers or scale

FIGURE 2.4 Rapids of Karun River in Ahvaz, Iran (By Mohammad Albaji).

FIGURE 2.5   Estuarine of a river; (https://www.sciencenewsforstudents.org/article/scientists-say-estuary).

models for the estuary configuration under consideration and reproducing in miniature the tide, flow, and fresh-water discharge over a sand bed, in which various lines of training walls can be successively inserted. The models should have ability for providing valuable indications of the respective effects and comparative merits of the different plans suggested (Figure 2.5).

EXERCISES

A. **Read each statement and decide whether it is true or false. Write "T" for true and "F" for false statements.**

TF1.   The first human actions in river engineering date back to a thousand years ago.

TF2.   The river size is dependent mainly on the extent of basin and the amount of basin rain.

TF3.   The purpose of canalization is to improve flow in flood conditions.

TF4.   Steep slope in the bed of a river is suitable for navigation.

TF5.   In lowering the shoals to provide a uniform depth in a river, the shoal material is determinant.

B. **Circle a, b, c, or d which best completes the following items.**

1.   .... is the process of human planned actions in the course, characteristics, or flow of a river for special purposes.
   a.   Channelization
   b.   River engineering
   c.   Canalization
   d.   Estuarine works

2. Part of a country that is confined to a watershed where rain falls to the lowest point is ....
   a. rapid section
   b. habitat
   c. basin of river
   d. bed of river

3. Installing .... in a large river and its tributaries at proper points can be useful in flood control.
   a. stream gauges
   b. locks
   c. obstructions
   d. shoals

4. .... purpose is to hold back the flow especially for navigation.
   a. Estuarine works
   b. Channelization
   c. Regulation works
   d. Canalization

5. Rivers have an increase in the extent of their drain basin when they move to the sea, owning to consecutive entry of ....
   a. rapids
   b. channels
   c. tributaries
   d. weirs

C. **Match the sentence halves in Column I with their appropriate halves in Column II. Insert the letters a, b, c ... in the parentheses provided. There are more sentence halves in Column II than required.**

Column I	Column II
1. River engineering	( ) **a.** is proportional to river fall.
2. Weir-keepers	( ) **b.** can be natural or artificial.
3. Obstructions	( ) **c.** needs a stable and navigable channel.
4. Rate of flow	( ) **d.** is protecting environment with some actions.
5. Erosion	( ) **e.** become able to open the movable weirs for flood passage timely by receiving information obtained from Stream gauges.
	( ) **f.** decreases by channelization.
	( ) **g.** includes two methods; channelization and canalization.
	( ) **h.** result in depth increasing and flow equalizing.

D. Cross out the word or words that make each statement false, and write the word or words that make each statement true in the blank.
1. When two rivers have an equal slope, the river that has a smaller size, has a faster flow.

   ....

2. Rivers flow velocity in lowest elevation of their basins, is fast, and torrents with a very variable flow.

   ....

3. The only method for diminution the effective fall is decreasing the length of channels by replacing winding sections with straight cuts.

   ....

4. The river level should be raised in summer by impounding the flow with stream gauges at intervals across the channel.

   ....

E. **Give answers to the following questions.**
1. What are the characteristics of rivers?
2. How stream gauges can be helpful in flood control?
3. What are the advantages and disadvantages of channelization?
4. For what purposes the interactions between river flow and tide should be modeled?

F. **For each word on the left, there are three meanings provided. Put a check mark (√) next to the choice which has the closest meaning to the word given.**

1. **Hydropower**	discharge	flow velocity	water power
2. **Headwater**	water elevation	river source	flood
3. **Gradient**	slope	river bed	flow rate
4. **Stream gauge**	pump station	stream level	gauging station
5. **Tributary**	river	affluent	pool

G. **Fill in the blanks with the appropriate words from the following list.**
river   bed   navigation   canalization   lowest   stage   obstructions   channelization   slope   flood   straight cuts   dry season   kflow   rate

River engineering is the process of targeted intervention of humans in the course, characteristics, or flow of a river for reaching certain purposes. Characteristics that human manipulate in rivers include size, basin, and .... of river. River engineering is generally categorized into two methods include channelization and canalization. In the .... the purpose is to improve flow, especially

in .... conditions by works such as replacing winding
sections with .... The elimination of ...., installing stream gauges. In the .... the purpose is to hold back
the flow, primarily for .... River engineering can be
beneficial for navigation conditions just in cases that large rivers
have a fair .... and adequate discharge at their
.... Steep slope in the .... is a problem for
navigation because of great variations in water level. In the
.... with very low discharge, it is impossible to provide
suitable water depth in the low-water channel.

H. **Read this passage and then circle a, b, c, or d which best
   completes the following items.**
   River engineering has not changed significantly compared to its
   early method over time. Past historical knowledge inside by intui-
   tion and experience is still the base of design, although some
   changes occurred in the amount, type, and detail of river data that
   are now available to engineers. In the last decade, new technologies
   such as the multi-beam hydrographic survey system for the collection of high-resolution bathymetry, and the acoustic Doppler
   current systems for the measurement of flow, have improved the
   ability of engineers to analyze rivers.
   Today, river engineering.....
   a. is exactly the same as the early method.
   b. have changed completely because of using new technologies.
   c. uses past historical knowledge combined with new technologies.
   d. is not advantageous same as past.

I. **Translate the following passage into your mother language.
   Write your translation in the space provided.**
   River engineering is a subset of civil engineering that designs and
   builds various structures to improve or repair rivers in order to
   respond to humans and environmental needs. Some of the river
   engineering structures include dikes, chevrons, bend way weirs, and
   banks. River engineering can improve navigation, reduce dredging,
   enhance or create new habitat, manage sediment, and control the
   erosion by deployment of modern-day, state of the art technologies
   in data collection and modeling. Sedimentation is one of the biggest
   natural human problems that can block the flow of rivers, clog water
   intakes for municipal water supply, disrupt the transport of com-
   modities by navigation, and destroy backwaters and wetlands.
   Erosion can endanger private property and infrastructure, cause
   major river cutoffs, and increase sedimentation.

J. **Copy the technical terms and expressions used in this lesson. Then find your mother language equivalents of those terms and expressions and write them in the spaces provided.**

Technical term	Mother language equivalent
....	....
....	....
....	....
....	....
....	....
....	....
....	....
....	....
....	....
....	....

BIBLIOGRAPHY

Aquapedia background, WATER EDUCATION FOUNDATION, 2020.

Congdon, J.C. (1971). Fish populations of channelized and unchannelized sections of the Chariton River, Missouri. In Schneberger, E., Funk, J.E. (eds.). *Stream Channelization–A Symposium*. North Central Division, American Fisheries Society. pp. 52–62.

Engineers Edge, LLC. *Fluid Volumetric Flow Rate Equation*. Engineers Edge. Retrieved 2016-12-01.

"Erosion". Encyclopædia Britannica. 2015-12-03. Archived from the original on 2015-12-21. Retrieved 2015-12-06.

Hinnant, L. (1970). Kissimmee River. In Marth, Del, Marth, Marty (eds.). *The Rivers of Florida*. Sarasota, FL: Pineapple Press. ISBN 0-910923-70-1.

"History of Hydropower | Department of Energy". energy.gov. Retrieved 4 May 2017.

https://en.wikipedia.org/wiki/Lock_(water_navigation)

https://en.wikipedia.org/wiki/Rapids

https://en.wikipedia.org/wiki/River_engineering
https://en.wikipedia.org/wiki/Stream_gauge
https://www.dictionary.com/browse/still-water
https://www.mvs-wc.usace.army.mil/arec/Basics.html
https://www.mvs-wc.usace.army.mil/arec/History.html

Jump up to:a b Neuendorf, K.K.E., J.P. Mehl, Jr., and J.A. Jackson, eds. (2005). *Glossary of Geology* (5th ed.). Alexandria, Virginia: American Geological Institute. 779 pp. ISBN 0-922152-76-4.

One or more of the preceding sentences incorporates text from a publication now in the public domain: Vernon-Harcourt, Leveson Francis (1911). River Engineering. In Chisholm, Hugh (ed.). *Encyclopædia Britannica. 23* (11th ed.). Cambridge University Press, pp. 374–385.

Pritchard, D.W. (1967). What is an estuary: physical viewpoint. In Lauf, G.H. (ed.). *Estuaries. A.A.A.S. Publ. 83.* Washington, DC. pp. 3–5. hdl:1969.3/24383.

Reddy, M.P.M. & Affholder, M. (2002). *Descriptive Physical Oceanography: State of the Art.* Taylor and Francis. p. 249. ISBN 90-5410-706-5. OCLC 223133263.

Rutecki, D., Nestler, E., Dellapenna, T., Pembroke, A. (2014). Understanding the Habitat Value and Function of Shoal/Ridge/Trough Complexes to Fish and Fisheries on the Atlantic and Gulf of Mexico Outer Continental Shelf. Draft Literature Synthesis for the U.S. Dept. of the Interior, Bureau of Ocean Energy Management. Contract # M12PS00031. Bureau of Ocean Energy Management, U.S. Department of the Interior. 116 pp.

The Ecological Effects of Channelization (The Impact of River Channelization). *Brooker, M.P.The Geographical Journal*, 1985, 151, 1, 63–69, The Royal Geographical Society (with the Institute of British Geographers).

Thomas, R. (2019). Fundamental of Ecology. *Marine Biology: An Ecological Approach (reprint ed.)* Waltham Abboy, Essex: Scientific e-Resources (published 2020). p. 86. ISBN 9781839474538. Archived from the original on 22 May 2020. Retrieved 8 March 2020. A habitat is an ecological or environmental area that is inhabited by a particular species of animal, plant, or other type of organism. The term typically refers to the zone in which the organism lives and where it can find food, shelter, protection and mates for reproduction.

"Tributary". PhysicalGeography.net, Michael Pidwirny & Scott Jones, 2009. Viewed 17 September 2012.

Zimmer, D.W., Bachmann, R.W. (1976). A Study of the Effects of Stream Channelization and Bank Stabilization on Warm Water Sport Fish in Iowa: the effects of long-reach channelization on habitat and invertebrate drift in some Iowa streams. Iowa Cooperative Fishery Research Unit, Iowa State University.

GLOSSARY OF TERMS

Erosion: In earth science, erosion is the action of surface processes such as water flow or wind that move soil, rock, or dissolved material from one place to another location on the Earth's crust.

Estuary: A partially enclosed coastal body of brackish water with one or more streams flowing into it, and connected to the sea.

Flow rate: The volume of fluid that passes per unit time.

Gradient: The ratio of decrease in elevation of a stream per horizontal distance that its unit is meters per kilometer or feet.

Habitat: In ecology, it is a type of natural environment in which a species of organism lives and can find food, refuge, protection, and mates for reproduction.

Headwater: The source of a river or stream that is located at the most distances from the outlet into the sea or where it is combined with another stream.

Hydropower: Power obtained from the energy of dropping or water fast move, which can be useful for some purposes.

Lock: A device used for raising and lowering vessels such as boats and ships between two different water levels in rivers and canal waterways.

Rapids: Sections of a river with a steep slope result in an increase in flow velocity and turbulence.

Still water: A part of a stream that is level or where the stream slope is so slight that no current is visible.

Stream gauges: A place where hydrologists or environmental scientists monitor and test terrestrial bodies of water.

Shoal: A natural submerged ridge, bank, or bar that is formed with sand or other unconsolidated material and rises from the water body bed to near the surface.

Tributary: A stream or river that discharges into a larger stream, main river, or lake.

Tide: Rise and fall of sea levels caused by gravity force of the Moon and the Sun and the rotation of the Earth.

3 Dams

READING FOR COMPREHENSION

A dam is a barrier that impounds water or underground streams. Dams generally serve the primary purpose of retaining water, while other structures such as floodgates or levees (also known as dikes) are used to manage or prevent water flow into specific land regions. Hydropower and pumped-storage hydroelectricity are often used in conjunction with dams to generate electricity. A dam can also be used to collect water or for storage of water which can be evenly distributed between locations (Figure 3.1).

TYPES OF DAMS

Dams can be formed by human agency, natural causes, or even by the intervention of wildlife such as beavers. Man-made dams are typically classified according to their size (height), intended purpose, or structure.

BY STRUCTURE

Based on structure and material used, dams are classified as timber dams, arch-gravity dams, embankment dams, or masonry dams, with several subtypes.

ARCH DAMS

In the arch dam, stability is obtained by a combination of arch and gravity action (Figure 3.2). If the upstream face is vertical the entire weight of the dam must be carried to the foundation by gravity, while the distribution of the normal hydrostatic pressure between vertical cantilever and arch action will depend upon the stiffness of the dam in a vertical and horizontal direction. When the upstream face is sloped the distribution is more complicated. The normal component of the weight of the arch ring may be taken by the arch action, while the normal hydrostatic pressure will be distributed as described above. For this type of dam, firm reliable supports at the abutments (either buttress or canyon side wall) are more important. The most desirable place for an arch dam is a narrow canyon with steep side walls composed of sound rock. The safety of an arch dam is dependent on the strength of the sidewall abutments, hence not only should the arch be well seated on the side walls but also the character of the rock should be carefully inspected (Figure 3.3).

Two types of single-arch dams are in use, namely the constant-angle and the constant-radius dam. The constant-radius type employs the same face radius at all elevations of the dam, which means that as the channel grows narrower toward the bottom of the dam the central angle subtended by the face of the dam becomes

DOI: 10.1201/9781003293507-3

FIGURE 3.1   Dez Dam, a double-arched concrete dam. Daz Lake in the background is impounded by the dam, Dezful, Iran (By Mohammad Albaji).

smaller. Jones Falls Dam, in Canada, is a constant radius dam. In a constant-angle dam, also known as a variable radius dam, this subtended angle is kept constant and the variations in distance between the abutments at various levels are taken care of by varying the radii. Constant-radius dams are much less common than constant-angle dams. Parker Dam is a constant-angle arch dam.

A similar type is the double-curvature or thin-shell dam. Wild horse Dam near Mountain City, Nevada in the United States is an example of the type. This method

FIGURE 3.2   An arch dam; (http://ussdams.com/ussdeducation/history.html).

FIGURE 3.3   Daniel-Johnson Dam, Quebec is the largest multiple-arch-and-buttress dam in the world; (http://seniorengineers.ca/csse/author/vannon).

of construction minimizes the amount of concrete necessary for construction but transmits large loads to the foundation and abutments. The appearance is similar to a single-arch dam but with a distinct vertical curvature to it as well lending it the vague appearance of a concave lens as viewed from downstream.

The multiple-arch dam consists of a number of single-arch dams with concrete buttresses as the supporting abutments, for example, the Daniel-Johnson Dam, Québec, Canada. The multiple-arch dam does not require as many buttresses as the hollow gravity type but requires a good rock foundation because the buttress loads are heavy.

GRAVITY DAMS

In a gravity dam, the force that holds the dam in place against the push from the water is Earth's gravity pulling down on the mass of the dam. The water presses laterally (downstream) on the dam, tending to overturn the dam by rotating about its toe (a point at the bottom downstream side of the dam). The dam's weight counteracts that force, tending to rotate the dam the other way about its toe. The designer ensures that the dam is heavy enough that gravity wins that contest. In engineering terms, that is true whenever the resultant of the forces of gravity and water pressure on the dam acts in a line that passes upstream of the toe of the dam (Figure 3.4).

Furthermore, the designer tries to shape the dam so if one were to consider the part of dam above any particular height to be a whole dam itself, that dam also would be held in place by gravity. i.e. there is no tension in the upstream face of the dam holding the top of the dam down. The designer does this because it is usually more practical to make a dam of material essentially just piled up than to make the material stick together against vertical tension.

Note that the shape that prevents tension in the upstream face also eliminates a balancing compression stress in the downstream face, providing additional economy.

The designer also ensures that the toe of the dam is sunk deep enough in the earth that it does not slide forward.

FIGURE 3.4   The Grand Coulee Dam is an example of a solid gravity dam; (http://www.flickr.com/photos/pdwhitney/2837620346/sizes/l).

For this type of dam, it is essential to have an impervious foundation with high *bearing* strength. When situated on a suitable site, a gravity dam can prove to be a better alternative to other types of dams. When built on a carefully studied foundation, the gravity dam probably represents the best developed example of dam building. Since the fear of flood is a strong motivator in many regions, gravity dams are being built in some instances where an arch dam would have been more economical.

Gravity dams are classified as "solid" or "hollow" and are generally made of either concrete or masonry. This is called "zoning." The core of the dam is zoned depending on the availability of locally available materials, foundation conditions and the material attributes. The solid form is the more widely used of the two, though the hollow dam is frequently more economical to construct. Gravity dams can also be classified as "overflow" (spillway) and "non-overflow." Grand Coulee Dam is a solid gravity dam and Itaipu Dam is a hollow gravity dam.

ARCH-GRAVITY DAMS

A gravity dam can be combined with an arch dam into an arch-gravity dam for areas with massive amounts of water flow but less material available for a pure gravity dam. The inwards compression of the dam by the water reduces the lateral (horizontal) force acting on the dam. Thus, the gravitation force required by the dam is lessened, i.e. the dam does not need to be so massive. This enables thinner dams and saves resources (Figure 3.5).

BARRAGES

A barrage dam is a special kind of dam that consists of a line of large gates that can be opened or closed to control the amount of water passing the dam. The gates are set between flanking piers which are responsible for supporting the water load. They are often used to control and stabilize water flow for irrigation systems (Figure 3.6).

FIGURE 3.5 The Marun Dam is an example of an arch-gravity, Iran.

FIGURE 3.6 The Pioneer River barrage, which regulates water flow along the Pioneer River; (http://wordpress.mrreid.org/2014/04/22/types-of-dam).

Barrages that are built at the mouth of rivers or lagoons to prevent tidal incursions or utilize the tidal flow for tidal power are known as tidal barrages.

EMBANKMENT DAMS

Embankment dams are made from compacted earth and have two main types, rock-fill and earth-fill dams. Embankment dams rely on their weight to hold back the force of water, like the gravity dams made from concrete.

ROCK-FILL DAMS

Rock-fill dams are embankments of compacted free-draining granular earth with an impervious zone. The earth utilized often contains a large percentage of large particles hence the term *rock-fill*. The impervious zone may be on the upstream face and made of masonry, concrete, plastic membrane, steel sheet piles, timber, or other material. The impervious zone may also be within the embankment in which case it is referred to as a *core*. In the instances where clay is utilized as the impervious material, the dam is referred to as a *composite* dam. To prevent internal erosion of clay into the rock-fill due to seepage forces, the core is separated using a filter. Filters are specifically graded soil designed to prevent the migration of fine grain soil particles. When suitable material is at hand, transportation is minimized leading to cost savings during construction. Rock-fill dams are resistant to damage from earthquakes. However, inadequate quality control during construction can lead to poor compaction and sand in the embankment which can lead to liquefaction of the rock-fill during an earthquake. Liquefaction potential can be reduced by keeping susceptible material from being saturated, and by providing adequate compaction during construction. An example of a rock-fill dam is New Melones Dam in California (Figure 3.7).

CONCRETE-FACE ROCK-FILL DAMS

A concrete-face rock-fill dam (CFRD) is a rock-fill dam with concrete slabs on its upstream face. This design offers the concrete slab as an impervious wall to prevent

FIGURE 3.7   The Gathright Dam in Virginia is a rock-fill embankment dam; (https://en. wikipedia.org/wiki/Dam).

leakage and also a structure without concern for uplift pressure. In addition, the CFRD design is flexible for topography, faster to construct, and less costly than earth-fill dams. The CFRD originated during the California Gold Rush in the 1860s when miners constructed rock-fill timber-face dams for sluice operations. The timber was later replaced by concrete as the design was applied to irrigation and power schemes. As CFRD designs grew in height during the 1960s, the fill was compacted and the slab's horizontal and vertical joints were replaced with improved vertical joints. In the last few decades, the design has become popular. Currently, the tallest CFRD in the world is the 233 m (764 ft) tall Shuibuya Dam in China which was completed in 2008.

Earth-Fill Dams

Earth-fill dams, also called *earthen dams*, *rolled-earth dams*, or simply *earth dams*, are constructed as a simple embankment of well-compacted earth. A *homogeneous* rolled-earth dam is entirely constructed of one type of material but may contain a drain layer to collect *seep* water. A *zoned-earth* dam has distinct parts or *zones* of dissimilar material, typically a locally plentiful *shell* with a watertight clay core. Modern zoned-earth embankments employ filter and drain zones to collect and re- move seep water and preserve the integrity of the downstream shell zone. An out- dated method of zoned earth dam construction utilized a hydraulic fill to produce a watertight core. *Rolled-earth* dams may also employ a watertight facing or core in the manner of a rock-fill dam. An interesting type of temporary earth dam occasionally used in high latitudes is the *frozen-core* dam, in which a coolant is circulated through pipes inside the dam to maintain a watertight region of permafrost within it.

Tarbela Dam is a large dam on the Indus River in Pakistan. It is located about 50 km (31 mi) northwest of Islamabad, and a height of 485 ft. (148 m) above the river bed and a reservoir size of 95 sq. mi (250 km^2) makes it the largest earth-filled dam in the world. The principal element of the project is an embankment 9,000 feet (2,700 meters) long with a maximum height of 465 feet (142 meters). The total volume of earth and rock used for the project is approximately 200 million cubic yards (152.8 million cu. Meters) which makes it the largest man-made structure in the world, except for the Great Chinese Wall which consumed somewhat more material.

Because earthen dams can be constructed from materials found on-site or nearby, they can be very cost-effective in regions where the cost of producing or bringing in concrete would be prohibitive.

ASPHALT-CONCRETE CORE

A third type of embankment dam is built with an asphalt concrete core. The majority of such dams are built with rock and/or gravel as the main fill material. Almost 100 dams of this design have now been built worldwide since the first such dam was completed in 1962. All asphalt-concrete core dams built so far have an excellent performance record. The type of asphalt used is a viscoelastic-plastic material that can adjust to the movements and deformations imposed on the embankment as a whole, and to settlements in the foundation. The flexible properties of the asphalt make such dams especially suited in earthquake regions.

BY SIZE

International standards (including International Commission on Large Dams, ICOLD) define *large dams* as higher than 15 meters and *major dams* as over 150 meters in height. The *Report of the World Commission on Dams* also includes in the *large* category, dams, such as barrages, which are between 5 and 15 meters high with a reservoir capacity of more than 3 million cubic meters.

The tallest dam in the world is the 300-meter-high Nurek Dam in Tajikistan.

BY USE

Saddle Dam

A saddle dam is an auxiliary dam constructed to confine the reservoir created by a primary dam either to permit a higher water elevation and storage or to limit the extent of a reservoir for increased efficiency. An auxiliary dam is constructed in a low spot or *saddle* through which the reservoir would otherwise escape. On occasion, a reservoir is contained by a similar structure called a dike to prevent inundation of nearby land. Dikes are commonly used for *reclamation* of arable land from a shallow lake. This is similar to a levee, which is a wall or embankment built along a river or stream to protect adjacent land from flooding.

Weir

A weir (also sometimes called an *overflow dam*) is a type of small overflow dam that is often used within a river channel to create an impoundment lake for water abstraction purposes and which can also be used for flow measurement or retardation.

Check Dam

A check dam is a small dam designed to reduce flow velocity and control soil erosion. Conversely, a *wing dam* is a structure that only partly restricts a waterway, creating a faster channel that resists the accumulation of sediment.

Dry Dam

A dry dam also known as a flood retarding structure is a dam designed to control flooding. It normally holds back no water and allows the channel to flow freely, except during periods of intense flow that would otherwise cause flooding downstream.

Diversionary Dam

A diversionary dam is a structure designed to divert all or a portion of the flow of a river from its natural course. The water may be redirected into a canal or tunnel for irrigation and/or hydroelectric power production.

Underground Dam

Underground dams are used to trap groundwater and store all or most of it below the surface for extended use in a localized area. In some cases, they are also built to prevent saltwater from intruding into a freshwater aquifer. Underground dams are typically constructed in areas where water resources are minimal and need to be efficiently stored, such as in deserts and on islands like the Fukuzato Dam in Okinawa, Japan. They are most common in northeastern Africa and the arid areas of Brazil while also being used in the southwestern United States, Mexico, India, Germany, Italy, Greece, France, and Japan.

There are two types of underground dams: a *sub-surface* and a *sand-storage* dam. A sub-surface dam is built across an aquifer or drainage route from an impervious layer (such as solid bedrock) up to just below the surface. They can be constructed of a variety of materials including bricks, stones, concrete, steel, or PVC. Once built, the water stored behind the dam raises the water table and is then extracted with wells. A sand-storage dam is a weir built in stages across a stream or wadi. It must be strong as floods will wash over its crest. Over time sand accumulates in layers behind the dam which helps store water and most importantly prevents evaporation. The stored water can be extracted with a well, through the dam body or by means of a drain pipe.

Tailings Dam

A tailings dam is typically an earth-fill embankment dam used to store tailings that are produced during mining operations after separating the valuable fraction from the uneconomic fraction of an ore. Conventional water retention dams can serve this purpose but due to cost, a tailings dam is more viable. Unlike water retention dams, a tailings dam is raised in succession throughout the life of the particular mine. Typically, a base or starter dam is constructed and as it fills with a mixture of tailings and water, it is raised. The material used to raise the dam can include the tailings (depending on their size) along with dirt.

There are three raised tailings dam designs, the *upstream*, *downstream*, and *centerline*, named according to the movement of the crest during raising. The specific design used is dependent upon topography, geology, climate, the type of tailings, and cost. An upstream tailings dam consists of trapezoidal embankments being constructed on top but toe to crest of another, moving the crest further upstream. This creates a relatively flat downstream side and a jagged upstream side which is supported by tailings slurry in the impoundment. The downstream design refers to the successive raising of the embankment that positions the fill and crest further downstream. A center-lined dam has sequential embankment dams constructed directly on top of another while fill is placed on the downstream side for support and slurry supports the upstream side.

Because tailings dams often store toxic chemicals from the mining process, they have an impervious liner to prevent seepage. Water/slurry levels in the tailings pond must be managed for stability and environmental purposes as well.

BY MATERIAL

Steel Dams

A steel dam is a type of dam briefly experimented with around the turn of the 19th–20th century which uses steel plating (at an angle) and load-bearing beams as the structure. Intended as permanent structures, steel dams were an (arguably failed) experiment to determine if a construction technique could be devised that was cheaper than masonry, concrete, or earthworks but sturdier than timber crib dams (Figure 3.8).

Timber Dams

Timber dams were widely used in the early part of the industrial revolution and in frontier areas due to ease and speed of construction. Rarely built in modern times because of relatively short lifespan and limited height to which they can be built, timber dams must be kept constantly wet in order to maintain their water retention properties and limit deterioration by rot, similar to a barrel. The locations where timber dams are most economical to build are those where timber is plentiful, cement is costly or difficult to transport, and either a low head diversion dam is required or longevity is not an issue. Timber dams were once numerous, especially in the North American west, but most have failed, been hidden under earth embankments or been replaced with entirely new structures. Two common variations of timber dams were the *crib* and the *plank* (Figure 3.9).

Timber crib dams were erected of heavy timbers or dressed logs in the manner of a log house and the interior filled with earth or rubble. The heavy crib structure supported the dam's face and the weight of the water. Splash dams were timber crib dams used to help float logs downstream in the late 19th and early 20th centuries.

Timber plank dams were more elegant structures that employed a variety of construction methods utilizing heavy timbers to support a water retaining arrangement of planks.

FIGURE 3.8 Red Ridge steel dam, born 1905, Michigan; (http://structurae.net/structures/redridge-steel-dam).

FIGURE 3.9 Green River Timber Crib Dam; (http://www.tug44.org/covered.bridges/green-river-covered-bridge).

FIGURE 3.10   A cofferdam during the construction of locks at the Montgomery Point Lock and Dam; (https://en.wikipedia.org/wiki/Dam).

OTHER TYPES

Cofferdams

A cofferdam is a (usually temporary) barrier constructed to exclude water from an area that is normally submerged. Made commonly of wood, concrete, or steel sheet piling, cofferdams are used to allow construction on the foundation of permanent dams, bridges, and similar structures. When the project is completed, the cofferdam may be demolished or removed. See also causeway and retaining wall. Common uses for cofferdams include construction and repair of offshore oil platforms. In such cases, the cofferdam is fabricated from sheet steel and welded into place under water. Air is pumped into the space, displacing the water allowing a dry work environment below the surface. Upon completion, the cofferdam is usually deconstructed unless the area requires continuous maintenance (Figure 3.10).

Beaver Dams

Beavers create dams primarily out of mud and sticks to flood a particular habitable area. By flooding a parcel of land, beavers can navigate below or near the surface and remain relatively well hidden or protected from predators. The flooded region also allows beavers access to food, especially during the winter.

Dam Failure

Dam failures are generally catastrophic if the structure is breached or significantly damaged. Routine deformation monitoring and monitoring of seepage from drains in and around larger dams are useful to anticipate any problems and permit remedial action to be taken before structural failure occurs. Most dams incorporate mechanisms to permit the reservoir to be lowered or even drained in the event of such

FIGURE 3.11 Failure of the Teton Dam near Rexburg, Idaho on Saturday June 6, 1976; (http://web.mst.edu/~rogersda/teton_dam/).

problems. Another solution can be rock grouting – pressure pumping Portland cement slurry into weak fractured rock (Figure 3.11).

During an armed conflict, a dam is to be considered as an "installation containing dangerous forces" due to the massive impact of a possible destruction on the civilian population and the environment. As such, it is protected by the rules of International Humanitarian Law (IHL) and shall not be made the object of attack if that may cause severe losses among the civilian population. To facilitate the identification, a protective sign consisting of three bright orange circles placed on the same axis is defined by the rules of IHL.

The main causes of dam failure include inadequate spillway capacity, piping through the embankment, foundation or abutments, spillway design error (South Fork Dam), geological instability caused by changes to water levels during filling or poor surveying (Vajont Dam, Malpasset, Testalinden Creek Dam), poor maintenance, especially of outlet pipes (Lawn Lake Dam, Val di Stava Dam collapse), extreme rainfall (Shakidor Dam), and human, computer or design error (Buffalo Creek Flood, Dale Dike Reservoir, Taum Sauk pumped storage plant).

A notable case of deliberate dam failure (prior to the above ruling) was the Royal Air Force "Dambusters" raid on Germany in World War II (codenamed *"Operation Chastise"*), in which three German dams were selected to be breached in order to have an impact on German infrastructure and manufacturing and power capabilities deriving from the Ruhr and Eder rivers. This raid later became the basis for several films.

Since 2007, the Dutch IJkdijk foundation is developing, with an open innovation model and early warning system for levee/dike failures. As a part of the development effort, full-scale dikes are destroyed in the IJkdijk field lab. The destruction process is monitored by sensor networks from an international group of companies and scientific institutions.

EXERCISES

A. **Read each statement and decide whether it is true or false. Write "T" for true and "F" for false statements.**

TF1. Man-made dams are typically classified according to their shape.

TF2. In a gravity dam, the force that holds the dam in place against the push from the water is Earth's gravity.

TF3. A dam is a barrier that impounds water or underground streams.

TF4. In the arch dam, stability is obtained only by gravity.

TF5. A barrage dam is a special kind of dam that consists of a line of large gates.

B. **Circle a, b, c, or d which best completes the following items.**

1. Gravity dams are classified as "solid" or "hollow" and are generally made of either ....
   a. stone and gravel
   b. soil and masonry
   c. concrete or masonry
   d. concrete and clay

2. .... can be combined with an arch dam into .... for areas with massive amounts of water flow but less material available for a pure gravity dam.
   a. An arch-gravity dam- a gravity dam
   b. An arch-gravity dam- an arch dam
   c. A gravity dam- an arch-gravity dam
   d. An arch dam- a gravity dam

3. .... dams are made from compacted earth and have two main types, rock-fill and .... dams.
   a. Embankment- earth-fill
   b. Earth - earth-fill
   c. Earth - stone-fill
   d. Stone-fill- earth-fill

4. ...... dams are .... of compacted free-draining granular earth with an impervious zone.
   a. Embankments - rock-fill
   b. Embankments- earth-fill
   c. Rock-fill- earth-fill
   d. Rock-fill- embankments

5. .... also called ...., rolled-earth dams, are constructed as a simple embankment of well-compacted earth.
   a. Earth-fill dams- earthen dams
   b. Earthen dams- earth-fill dams
   c. Earth dams- earth-fill dams
   d. a, b, & c

C. **Match the sentence halves in Column I with their appropriate halves in Column II. Insert the letters a, b, c ... in the parentheses provided. There are more sentence halves in Column II than required.**

Column I	Column II
1. Two types of single-arch dams are in use,	( ) **a.** used to trap groundwater and store all or most of it below the surface for extended use in a localized area.
2. A concrete-face rock-fill dam is	( ) **b.** a flood retarding structure, is a dam designed to control flooding.
3. A check dam is	( ) **c.** namely the constant-angle and the constant-radius dam.
4. Underground dam is	( ) **d.** a structure designed to divert all or a portion of the flow of a river from its natural course.
5. A dry dam is also known as	( ) **e.** a barrier constructed to exclude water from an area that is normally submerged.
	( ) **f.** is a weir built in stages across a stream or wadi. It must be strong as floods will wash over its crest.
	( ) **g.** a rock-fill dam with concrete slabs on its upstream face.
	( ) **h.** a small dam designed to reduce flow velocity and control soil erosion.

D. **Cross out the word or words that make each statement false, and write the word or words that make each statement true in the blank.**

1. A Beaver dam is typically an earth-fill embankment dam used to store tailings—which are produced during mining operations after separating the valuable fraction from the uneconomic fraction of an ore.

    ....

2. Steel dams were widely used in the early part of the industrial revolution and in frontier areas due to ease and speed of construction.

    ....

3. A cofferdam is an auxiliary dam constructed to confine the reservoir created by a primary dam either to permit a higher water elevation and storage or to limit the extent of a reservoir for increased efficiency.

    ....

4. A check dam is a type of small overflow dam that is often used within a river channel to create an impoundment lake for water abstraction purposes and which can also be used for flow measurement or retardation.

    ....

E. **Give answers to the following questions.**
   1. What is a dam?
   2. What are the types of dams (by structures)?
   3. What are the embankment dams?

F. **For each word on the left, there are three meanings provided. Put a check mark (√) next to the choice which has the closest meaning to the word given.**

1. **Earth-fill dams**	rolled-earth dams	soil dams	gravel dams
2. **Weir**	beaver dam	cofferdam	overflow dam
3. **Underground dams**	undersurface dam	clay-storage dam	sand-storage dam
4. **Dry dam**	embankment dam	flood retarding structure	earth dam
5. **Embankment dams**	earth-fill dams	soil dams	clay dams

G. **Fill in the blanks with the appropriate words from the following list.**

arch dam   world Bank   country   development   capacity
consortium   Khuzestan   development   hydroelectric
power station

The Dez Dam is an .... on the Dez River in the southwestern province of ...., Iran. It is about 26 km (16 mi) north of Andimeshk. It was built between 1959 and 1963 by an Italian .... and is owned by the Khuzestan Water and Power Authority. The dam is 203 meters (666 ft.) high, making it one of the highest in the ...., and has a reservoir .... of 3,340,000,000 m^3 (2,710,000 acre·ft). At the time of construction, the Dez Dam was Iran's biggest .... project. The primary purpose of the dam is .... power production and irrigation. It has an associated 520 MW ....and its reservoir helps irrigate up to 80,500 ha (199,000 acres) of farmland. US$42 million of the cost to construct the dam came from the ....

FIGURE 3.12 The Dez Dam, Dezful, Iran; (https://en.wikipedia.org/wiki/Dez_Dam).

H. **Read this passage and then circle a, b, c, or d which best completes the following items.**

One of the engineering wonders of the ancient world was the Great Dam of Marib in Yemen. Initiated somewhere between 1750 and 1700 BC, it was made of packed earth – triangular in cross-section, 580 m in length and originally 4 meters high – running between two groups of rocks on either side, to which it was linked by substantial stonework. Repairs were carried out during various periods, most important around 750 BC, and 250 years later the dam height was increased to 7 meters. After the end of the Kingdom of Saba, the dam fell under the control of the Ḥimyarites (~115 BC) who undertook further improvements, creating a structure 14 meters high, with five spillway channels, two masonry-reinforced sluices, a settling pond, and a 1000 meter canal to a distribution tank. These extensive works were not actually finalized until 325 AD and allowed the irrigation of 25,000 acres (100 km^2) (Figure 3.13).

FIGURE 3.13 The current Great Dam of Marib in 1986, Yemen; (https://en.wikipedia.org/wiki/Marib_Dam).

The main purpose of this passage is ....

a. introduction of types of dams.
b. introduction of ancient dams.
c. introduction of heritage.
d. introduction of Kingdom of Saba.

I. **Translate the following passage into your mother language. Write your translation in the space provided.**
Dams can also be created by natural geological forces. Volcanic dams are formed when lava flows, often basaltic, intercept the path of a stream or lake outlet, resulting in the creation of a natural impoundment. An example would be the eruptions of the Uinkaret volcanic field about 1.8 million–10,000 years ago, which created lava dams on the Colorado River in northern Arizona in the United States. The largest such lake grew to about 800 kilometers (500 mi) in length before the failure of its dam. Glacial activity can also form natural dams, such as the damming of the Clark Fork in Montana by the Cordilleran Ice Sheet, which formed the 7,780 km^2 (3,000 sq mi) Glacial Lake Missoula near the end of the last Ice Age. Moraine deposits left behind by glaciers can also dam rivers to form lakes, such as at Flathead Lake, also in Montana.

Natural disasters such as earthquakes and landslides frequently create landslide dams in mountainous regions with unstable local geology. Historical examples include the Usoi Dam in Tajikistan, which blocks the Murghab River to create Sarez Lake. At 560 m (1,840 ft.) high, it is the tallest dam in the world, including both natural and man-made dams. A more recent example would be the

FIGURE 3.14   The edge of The Barrier in British Columbia, Canada; (https://en.wikipedia.org/wiki/The_Barrier).

creation of Attabad Lake by a landslide on Pakistan's Hunza River. Natural dams often pose significant hazards to human settlements and infrastructure. The resulting lakes often flood inhabited areas, while a catastrophic failure of the dam could cause even greater damage such as the failure of western Wyoming's Gros Ventre landslide dam in 1927, which wiped out the town of Kelly and resulted in the deaths of six people.

J. Copy the technical terms and expressions used in this lesson. Then find your mother language equivalents of those terms and expressions and write them in the spaces provided.

Technical term	Mother language equivalent
....	....
....	....
....	....
....	....
....	....
....	....
....	....
....	....
....	....

BIBLIOGRAPHY

Adam, L. (2006). *Wind, Water, Work: Ancient and Medieval Milling Technology*, p. 62. BRILL, ISBN 90-04-14649-0.

Agoramoorthy, G., Chaudhary, S., & Minna, J. (2011). *The Check-Dam Route to Mitigate India's Water Shortages.* Law Library – University of New Mexico. http://lawlibrary. unm.edu/nrj/48/3/03_agoramoorthy_indian.pdf. Retrieved 8 November 2011.

Asphalt concrete cores for embankment dams. *International Water Power and Dam Construction.* http://www.waterpowermagazine.com/story.asp?storyCode=472. Retrieved 3 April 2011.

Blackwater Dam. *US Army Corps of Engineers.* http://www.nae.usace.army.mil/recreati/ bwd/bwdfc.htm. "The principal objective of the dam and reservoir is to protect downstream communities".

Dams and Development: An Overview. 16 November 2000. http://www.dams.org/report/ wcd_overview.htm. Retrieved 24 October 2010. "Box 1. What is a large dam?"

Dams the latest culprit in global warming. *Times of India.* 8 August 2012. http://timesofindia. indiatimes.com/home/environment/global-warming/Dams-the-latest-culprit-in-global-warming/articleshow/15403985.cms. Retrieved 9 August 2012.

Environmental issues and management of waste in energy and mineral production: proceedings of the Sixth International Conference on Environmental Issues and Management of Waste in Energy and Mineral Production: SWEMP 2000; Calgary, Alberta, Canada, May 30 – June 2, 2000. Rotterdam [u.a.]: Balkema. 2000. pp. 257–260. ISBN 90-5809-085-X. http://books.google.com/?id=PqiYy538JFUC& pg=PA257&dq=tailings+dam#v=onepage&q=tailings%20dam&f=false

Guinness Book of Records 1997 Pages 108–109 ISBN 0-85112-693-6.

Günther Garbrecht: "Wasserspeicher (Talsperren) in der Antike". *Antike Welt,* 2nd special edition: *Antiker Wasserbau* (1986), pp. 51–64 (52).

Helms, S.W. (1975). Jawa Excavations. Third Preliminary Report, Levant 1977. http://books.google.com/?id=qG9Bux3RYWMC&pg=PA9&dq=tailings+dam#v=onepage& q=tailings%20dam&f=false. Retrieved 10 August 2011.

http://seniorengineers.ca/csse/author/vannon
http://structurae.net/structures/redridge-steel-dam
http://ussdams.com/ussdeducation/history.html
http://web.archive.org/web/20070206130842/
http://web.mst.edu/~rogersda/teton_dam/
http://wordpress.mrreid.org/2014/04/22/types-of-dam
http://www.britishdams.org/about_dams/gravity.htm
http://www.chincold.org.cn/news/li080321-17-shuibuya.pdf. Retrieved 23 August 2011.
http://www.flickr.com/photos/pdwhitney/2837620346/sizes/l
http://www.hindunet.org/saraswati/traditionwater.pdf.
http://www.pbs.org/wgbh/buildingbig/dam/basics.html#arch. Retrieved 7 January 2007.
http://www.tug44.org/covered.bridges/green-river-covered-bridge
https://en.wikipedia.org/wiki/Dam
https://en.wikipedia.org/wiki/Dez_Dam
https://en.wikipedia.org/wiki/Marib_Dam
https://en.wikipedia.org/wiki/The_Barrier
Is it Worth a Dam?". Environmental Health Perspectives Volume 105, Number 10, October 1997 (Archived 17 May 2006 at the Wayback Machine).
Kazakhstan. *Land and Water Development Division*. (1998). "Construction of a dam (Berg Strait) to stabilize and increase the level of the northern part of the Aral Sea."
Methodology and Technical Notes. *Watersheds of the World*. Archived from the original on 4 (2007). http://web.archive.org/web/20070704103642/http://www.iucn.org/themes/wani/eatlas/html/technotes.html. Retrieved 1 August 2007.
Mohamed Bazza (28-30). "Overview of the history of water resources and irrigation management in the near east region" (PDF). http://www.fao.org/world/Regional/RNE/morelinks/Publications/English/HYSTORY-OF-WATER-RESOURCES.pdf. Retrieved 1 August 2007.
Needham, J. (1986). *Science and Civilization in China: Volume 4, Part 3*. Taipei: Caves Books, Ltd.
Neves, edited by E. Maranha das (1991). *Advances in rockfill structures*. Dordrecht: Kluwer Academic. p. 341. ISBN 0-7923-1267-8. http://books.google.com/?id=USEyV8y9ZFQC&pg=PA341&dq=concrete+face+rock+fill+dams#v=onepage&q=concrete%20face%20rock%20fill%20dams&f=false.
Onder, H., & Yilmaz, M. (2005). Underground Dams – A Tool of Sustainable Development and Management of Ground Resources. *European Water*, 35–45. http://www.ewra.net/ew/pdf/EW_2005_11-12_05.pdf. Retrieved 7 May 2012.
Properties of Tailings Dams. *NBK Institute of Mining Engineering*. http://www.mining.ubc.ca/faculty/meech/MINE290/Tailings%20Dam%20Construction%20Methods.pdf. Retrieved 10 August 2011.
Renewable Global Status Report 2006 Update, REN21, published 2006, accessed 16 May 2007.
Routledge Hill, D. (1996a), "Engineering", p. 759, in Rashed, Roshdi; Morelon, Régis (1996). Encyclopedia of the History of Arabic Science. Routledge. pp. 751–795. ISBN 0-415-12410-7.
Routledge Hill, D. (1996b). *A history of engineering in classical and medieval times*. Routledge. pp. 56–58. ISBN 0-415-15291-7.
Singh, V., & Ram Narayan Yadava P., (2003). *Water Resources System Operation: Proceedings of the International Conference on Water and Environment*. Allied Publishers. p. 508. ISBN 81-7764-548-X. http://books.google.com/?id=Bge-0XX6ip8C&pg=PA508&dq=kallanai#PPA508,M1.
The American Heritage Dictionary of the English Language, Fourth Edition.
The Dez Multipurpose Dam Project in Iran. San Jose State University. Retrieved 13 January 2015.

The Impact of Agricultural Development on Aquatic Systems and its Effect on the Epidemiology of Schistosomes in Rhodesia" (PDF). IUCN. "Recently, agricultural development has concentrated on soil and water conservation and resulted in the construction of a multitude of dams of various capacities which tend to stabilize water flow in rivers and provide a significant amount of permanent and stable bodies of water."

The reservoirs of Dholavira. The South Asia Trust. December 2008. http://himalmag. com/component/content/article/44/1062-The-reservoirs-of-Dholavira.html. Retrieved 27 February 2011.

Three Gorges dam wall completed. *China-embassy.* 20 May 2006. http://www.china-embassy.org/eng/zt/sxgc/t36502.htm. Retrieved 21 May 2006.

Water Reservoirs behind Rising Greenhouse Gases. *French Tribune.* 9 August 2012. http:// frenchtribune.com/teneur/1212763-water-reservoirs-behind-rising-greenhouse-gases. Retrieved 9 August 2012.

World Commission on Dams Report. *Internationalrivers.org.* 29 February 2008. http:// internationalrivers.org/en/way-forward/world-commission-dams/world-commission-dams-framework-brief-introduction. Retrieved 16 August 2012.

Yilmaz, M. (2003). *Control of Groundwater by Underground Dams.* The Middle East Technical University. http://etd.lib.metu.edu.tr/upload/1259621/index.pdf. Retrieved 7 May 2012.

GLOSSARY OF TERMS

Arch dam: A dam that is curved in the horizontal plane and usually built of concrete, in which the horizontal thrust is taken by abutments in the sides of a valley. Arch dams must be built on solid rock, as a yielding material would cause a failure

Arch-gravity dam: A gravity dam can be combined with an arch dam into an arch-gravity dam for areas with massive amounts of water flow but less material available for a pure gravity dam.

Barrage: A Barrage dam is a special kind of dam that consists of a line of large gates that can be opened or closed to control the amount of water passing the dam.

Check dam: A check dam is a small dam designed to reduce flow velocity and control soil erosion.

Cofferdams: A cofferdam is a (usually temporary) barrier constructed to exclude water from an area that is normally submerged.

Dry dam: A dry dam also known as a flood retarding structure is a dam designed to control flooding.

Earth-fill dam: Earth fill dam, also called Earth Dam, or Embankment Dam, dam built up by compacting successive layers of earth, using the most impervious materials to form a core and placing more permeable substances on the upstream and downstream sides.

Embankment dam: A dam structure constructed of fill material, usually earth or rock, placed with sloping sides and usually with a length greater than its height.

Dam: A dam is a barrier that impounds water or underground streams.

Dike: An impermeable linear structure for the containment of over-bank flow. Dikes are similar to levees but generally much shorter.

Diversionary dam: A diversionary dam is a structure designed to divert all or a portion of the flow of a river from its natural course. The water may be redirected into a canal or tunnel for irrigation and/or hydroelectric power production.

Gravity dam: A dam that depends on its weight for stability.

Rock-fill dam: A rock-fill dam is a type of embankment dam that comprises primarily compacted rock materials.

Saddle dam: A saddle dam is an auxiliary dam constructed to confine the reservoir created by a primary dam either to permit a higher water elevation and storage or to limit the extent of a reservoir for increased efficiency.

Spillway: A waterway in or about a dam or other hydraulic structure for the passage of excess water.

Tailings dam: A tailings dam is typically an earth-fill embankment dam used to store tailings – which are produced during mining operations after separating the valuable fraction from the uneconomic fraction of an ore.

Underground dam: Underground dams are used to trap groundwater and store all or most of it below the surface for extended use in a localized area.

4 Hydroinformatics

READING FOR COMPREHENSION

Hydroinformatics is a branch of informatics, which concentrates on the application of information and communications technologies (ICTs) to meet the equitable and efficient management and use of water for a wide variety of purposes. Hydroinformatics, as a relatively new area of research, is the interdisciplinary domain between water science, data science, computer science, environmental science, and many other disciplines.

HYDROINFORMATICS HISTORY

While the concepts of computational hydraulics and hydrological modeling have been existed since at least the late 1950, in the late 1980s, hydrological numerical modeling, data collection, and data processing were beginning to expand and synchronize; more variables considered in water flow studies, recordings and samplings scope expanded, and networked computer systems became mightier. Thus, numerical modeling of water expanded its scope from the flows modeling to a socio-technical field that contributed users to address their water-related problems. This evolution led to the introduction of the term hydroinformatics by Professor M.B. Abbott in his book published in 1991. Inspired by the concept of computational hydraulics, he defined hydroinformatics as a combination of computational hydraulics and artificial intelligence.

Over the past three decades, progressive international activities on hydroinformatics topic have taken place. The International Association for Hydraulic Research (IAHR) started international activities on hydroinformatics field in associations formed with the establishment of a hydroinformatics section for Hydraulic Research (IAHR) in 1993. Then the International Water Association (IWA) established a Specialist Group on Hydroinformatics in 1998. Finally, the International Association of Hydrological Sciences (IAHS) joined IAHR and IWA forming the IAHR IWA-IAHS Joint Committee on Hydroinformatics.

The International Conference on Hydroinformatics (HIC) was held in September 1994, for the first time in IHE Delft Institute for Water Education, Netherlands. Since then, until 2021, the HIC has been held every two years, 14 times, around the world (e.g. Copenhagen, Denmark; Cardiff, Wales; Nice, France; Tianjin, China; Incheon, South Korea; Mexico City, Mexico).

The first education course on hydroinformatics topic established by UNESCO-IHE Institute for Water Education in 1991 as a Master's program. In addition, the University of Illinois conducted an experimental graduate-level course on this topic in 2003. Today, in many countries around the world, universities offer graduate and

DOI: 10.1201/9781003293507-4

postgraduate courses in hydroinformatics topics such as Utah State University and Brigham Young University in USA, and Newcastle University in England.

How the Hydroinformatics Work

The increasing global water crisis partly reflects the failure of human society to be aware of the problem and its possible solutions. Peoples look at this crisis from different perspectives. Some see it as a conflict of interest between different stakeholders of water resources, such as between industrialists and agriculturalists. From another group perspective, water crisis is about the management of climate change problems and its effects include drought and desertification or more frequent and severe flooding in rivers and from coastal waters. For another group, it is handling the problem of existing water resources pollution, whether originating from point sources (industry and urban wastewater) or non-point sources (agriculture and varying land uses). Considering all the dimensions mentioned, water scientists have realized that water management must involve an integrated view of a number of distinct systems, instead of dealing separately, to meet the demands of a wide range of stakeholders. As a result, it is necessary to collaborate with specialists from several other disciplines.

Many integrated systems have very complex interactions, and the normal way of trying to overcome the complexity is to form a model of them. The modeling and analysis of complex water-based systems depend almost entirely on the development of computers and digital technologies because our minds are unable to do this due to complexity and calculations. Hydroinformatics can implement this modeling. Modeling has two main purposes; the first is to make a better understanding of the real-world domain performance through reproducing past performances and understanding why they occurred, and second is performance predictions (Figure 4.1).

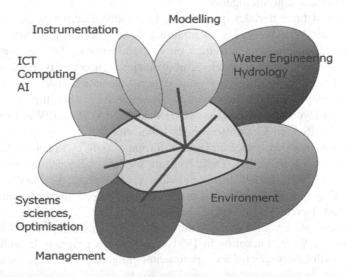

FIGURE 4.1 Position of hydroinformatics; (Price and Solomatine, (2009)).

Hydroinformatics deals with three main modeling paradigms including physically-based modeling, data-driven modeling, and agent-based modeling.

PHYSICALLY-BASED MODELING

Physically-based modeling (also called process-based, numerical, or simulation modeling) is based on a knowledge of the physics (physics of the flow of water), chemistry (chemistry of the associated substances), or biology (biology of the ecology in the aquatic environment) that is somehow in the form of software. One of the most known examples in water science is the one-dimensional continuity and momentum equations for open channel flow which is known as Saint-Venant equation. Since these equations have no analytical solution (except under very limited conditions), the numerical solution must be apply. The numerical solutions generate appropriate numerical equations based on a finite difference, finite element, or finite volume approach using a discretization of the solution space. The generated numerical equations are solved using particular algorithms that are coded in a particular software language and run on a computer. The software sets a direct link between the input to model and the corresponding output.

Physically-based models:

- Are usually deterministic in that there is a unique output for a given input, provided the complete input data.
- Can generate large amount of information about what occurs beyond the data (input or output) boundaries.
- Can made modifications in structural objects to evaluate the system performance in various scenarios involving structural change.
- After calibration, confirmation can be used with some degree of confidence for a range of input data covered by the calibration, or even for a limited degree of extrapolation due to the encapsulated physics. The confidence degree of the model for extrapolation depends on the quality of the structural data and possibly on the definition of some critical processes.

DATA-DRIVEN DODELING

Data-driven modeling (DDM), empirical modeling, is based on the direct analysis of the characteristics data about the understudy system, in particular, detect the connections between the system state variables (input, internal, and output variables) without clear knowledge (with only a limited number of assumptions) about the physical behavior of the system.

Data-driven modeling is very different from physically-based modeling, although the goal of both is connecting one data set (output) to another corresponding set (input). The main idea is to work with the given data only at the domain boundary and to find a form of relationship(s) that best connects the specific data sets. The characteristic feature of data-driven modeling is learning from available data, which involves the unknown mappings or dependencies between system inputs and outputs.

Using the data, we identify the known samples that are a combination of inputs and corresponding outputs. Subsequently, a dependency is discovered which can be used to predict or effectively deduce the future outputs of the system from known input values. There is a mathematical equation in data-driven modeling which derived from the analysis of time series data, not from physical processes, such as the unit hydrograph method, linear regression, and ARIMA models. Data-driven modeling is undergoing fast development and capabilities expansion due to recent developments in computational intelligence, especially in the field of machine learning. The contemporary methods go far beyond those applied in common empirical modeling in hydraulic engineering and hydrology.

There is a various range of data-driven modeling techniques including artificial neural networks, fuzzy rule-based systems, nearest neighbor, decision/model trees, support vector machines, chaos theory and non-linear dynamics, genetic programming and evolutionary regression, genetic algorithms in model optimization. The most widely used technique is artificial neural networks, which are described below.

ARTIFICIAL NEURAL NETWORKS

Artificial neural networks (ANNs), simply called neural networks (NNs), are computational models inspired by the biological neural networks that are based on human and animal brain information processing method, although do not try to be biologically realistic in detail. ANNs are one of the well-established technologies in machine learning and a major data-driven modeling technique.

An ANN consists of a large number of simple processing elements called neurons, nodes, or units. Typically, there are three layers of neurons in an ANN that formed for modeling the connectivity between a time series input and a corresponding time series output; an input layer with a number of specific inputs, a hidden layer containing another number of neurons, and an output layer with one or more neurons.

Signals travel from the first layer (input layer) to the last layer (output layer), probably after passing through the layers several times. Different layers may perform different transformations on their inputs.

Each neuron has an internal state named its activity level or activation, which is a function of all the inputs it has received. Based on the result of its activation, the neuron sends one signal through connections (similar to the synapses in a biological brain) to several other neurons each time. An artificial neuron that receives a signal processes it and then can send it to connected neurons. The signal in the connection is a real number, and the output of each neuron is calculated by some non-linear functions from the sum of its inputs. Neurons and edges (connections) usually have a weight that adjusts as learning proceeds. The weight increases or decreases the signal strength in a connection. Neurons may have a threshold such that a signal is sent only if the aggregate signal goes over that threshold. Neural network models are developed by training the network to identify the inherent relationships and processes of data. The hidden layer is the essential component that allows the neural

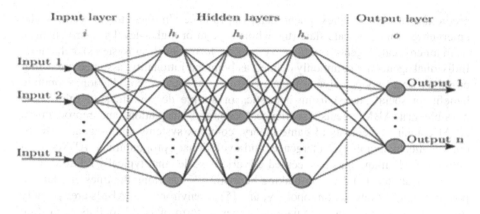

FIGURE 4.2   The schematic of the artificial neural network.

network to learn the relationships in the data. Learning process consists essentially of altering the synaptic connections strength (weights) between the neural cells.

In other words, each neuron receives inputs either externally or from other neurons, and then passes them through an activation or transfer function such as a logistic or sigmoid curve. Data enter the network through the input units arranged in the input layer, then fed forward through consecutive layers involving the hidden layer in the middle to exit from the output layer on the right. In this process, some knowledge about the hydrological system is important because the inputs can be any combination of variables that are thought to be important for predicting the output.

The validation data set is the second set of data given to the network for evaluation during training. If this approach is not used, the network will display the training data set well but then cannot generalize to an unseen data set or a testing data set. The main act to contribute promote generalization to unseen data is to make sure that the training data set involves a representative sample of the entire data behavior. This assurance can be met by ensuring that all three data sets (training, validation, and test) have similar statistical properties (Figure 4.2).

AGENT-BASED MODELING

Agent-based modeling (ABM) is computational modeling for simulating the actions and interactions of autonomous entities (agents) in a complex adaptive system (CAS) in order to understand the behavior of a system and what governs its outcomes. Despite the distinction, agent-based modeling is related to the concept of multi-agent systems or multi-agent simulation because the ABM purpose is to look for explanation of the collective behavior of agents following simple rules (especially in natural systems), instead of designing agents and solving specific practical or engineering problems. ABMs are a kind of microscale models that simulate the simultaneous operations and interactions of multiple agents to recreation and prediction of the complex phenomena appearance. In fact, the process is the emergence of macro-scale state (the whole system or higher-level system) changes from micro-scale agents (parts of system or lower-level subsystems) behaviors. Individual agents are commonly defined as bounded rationality, assumed are acting based on what they perceive as their own interests (such as reproduction, economic benefit, or social status) using heuristics or simple decision-making rules. It is possible that ABM agents experience "learning," adaptation, and reproduction. ABM combines elements of game theory, complex systems, multi-agent systems, computational sociology, emergence, and evolutionary programming. ABMs often consist of: (1) many agents specified at various scales that typically referred to as agent-granularity; (2) decision-making heuristics; (3) learning the rules or adaptive processes; (4) an interaction topology; and (5) an environment. ABMs are typically implemented as computer simulations, either in form of ABM toolkits or custom software. ABMs are used in many scientific fields including biology, ecology, and social science. ABMs are also called individual-based models (IBMs), particularly in ecology.

EXERCISES

A. **Read each statement and decide whether it is true or false. Write "T" for true and "F" for false statements.**

TF1. Hydroinformatics is an interdisciplinary discipline that employs a various field studies.

TF2. The computational hydraulics concepts have been introduced in the late 1980s.

TF3. It is not possible to have an integrated view on water management because there are different views on this problem.

TF4. If the input data is complete, physically-based models usually provide a unique output for a given input.

TF5. In artificial neural networks, the network is able to generalize the training data set results to an unseen data set or a testing data set without validation stage.

B. **Circle a, b, c, or d which best completes the following items.**

1. Professor M.B. Abbott defined hydroinformatics as a combination of.....…..............
   a. computational hydraulics and hydrology.
   b. water science and data science.
   c. social and technical fields.
   d. computational hydraulics and artificial intelligence.

2. The first main purpose of modeling is ....
   a. to make accurate prediction of performance.
   b. integrated water management.
   c. to achieve better comprehension of the real-world domain performance.
   d. to save time and money.

3. .... is not one of the data-driven modeling techniques.
   a. Fuzzy rule-based systems
   b. Decision-making heuristics
   c. Support vector machines
   d. Nearest neighbor
4. The output of each neuron is calculated by .... from the sum of its inputs.
   a. linear functions
   b. sigmoid curve
   c. linear regression
   d. non-linear functions
5. In an artificial neural network, the .... is the essential component to learn the relationships in the data.
   a. hidden layer
   b. input layer
   c. output layer
   d. last layer

C. **Match the sentence halves in Column I with their appropriate halves in Column II. Insert the letters a, b, c ... in the parentheses provided. There are more sentence halves in Column II than required.**

Column I	Column II
1. Computational hydraulics	() **a.** study of the computer algorithms which makes systems able to automatically learn and improve from experience without being explicitly programmed.
2. Heuristic	() **b.** the elements arrangement in a communication network.
3. Machine learning	() **c.** the path of signal transmission from one neuron to another neuron or to the target effector cell.
4. Network topology	() **d.** any prescribed way of assigning to each object in one set a particular object in another set.
5. Synapse	() **e.** a field of study that deals with the imitation of human intelligence processes
	() **f.** a problem-solving technique that is sufficient to achieve an immediate goal or approximation.
	() **g.** a science that combines different disciplines with the aim of simulating various physical processes by computers.
	() **h.** the combination of two or more academic disciplines into one activity.

D. **Cross out the word or words that make each statement false, and write the word or words that make each statement true in the blank.**

1. The International Conference on Hydroinformatics (HIC) was held for the first time at the Copenhagen Institute for Water Education, Denmark in 1996.

   ....

2. Modeling and analyzing complex water-based systems is almost independent of the development of computers and digital technologies because the human mind can do the same.

   ....

3. Signals pass through different layers that perform the same transformations on their inputs.

   ....

4. The purpose of agent-based modeling is to look for design agents and solve specific practical problems rather than explanation of the collective behavior of agents following simple rules.

   ....

E. **Give answers to the following questions.**

1. What is hydroinformatics?
2. What are the physically-based models characteristics?
3. What are data-driven modeling techniques?
4. How does an agent-based model work?

F. **For each word on the left, there are three meanings provided. Put a check mark (√) next to the choice which has the closest meaning to the word given.**

1. **Process-based models**	data-driven models	ARIMA models	multiple models
2. **Neurons**	edges	units	networks
3. **Edges**	neurons connections	signals	domain boundary
4. **Agents**	weights	synapses	autonomous entities
5. **Agent-based models**	empirical models	individual-based models	simulation models

G. **Fill in the blanks with the appropriate words from the following list.**

weights	nodes	numerical	mappings
activation	microscale	Hydroinformatics	desertification

The global water crisis and its related problems such as drought and .... require an integrated management using interdisciplinary methods. Considering the complexity of integrated systems and the limitations of the human mind, we need digital technologies to model complex water-based systems. ...., as a science that can meet this need, has focused on three main modeling paradigms including physically-based modeling, data-driven modeling, and agent-based modeling. Physically-based modeling, also called .... modeling, is based on a knowledge of physics, chemistry, or biology, which is formed in software. These models can generate large amount of information about what occurs beyond the data boundaries. Data-driven modeling is based on the learning from available data, which involves the unknown .... or dependencies between system inputs and outputs. Artificial neural networks are one of the major data-driven modeling technique, in which consists of a large number of simple processing elements named .... Neurons have an internal state called .... Learning process consists of altering the .... between the neural cells. Agent-based models are a kind of .... models that simulate the simultaneous operations and interactions of multiple agents to recreation and prediction of the complex phenomena appearance.

H. **Read this passage and then circle a, b, c, or d which best completes the following items.**
Every model is an approximation to reality, so a main issue in modeling is the uncertainty. considering that measurements, particularly of flows, can have an uncertainty of 20% or even more, therefore the models that are built using such data will have uncertainties of at least a similar order. The decision-maker needs to know how safe and reliable the results from a model are in affecting the decision made. Uncertainty estimation is still rarely used, however, considerable effort is now put into reduce the models error, and there are many studies in relation to water-related issues, that estimated the model uncertainty.

There are some reasons for reluctant to the uncertainty estimation in hydrological and hydraulic modeling including:
- Uncertainty analysis is too subjective, and difficult to perform.
- Uncertainty analysis cannot be incorporated into the decision-making process, and does not really important in making the final decision.
- Uncertainty analysis is not necessary considering physically realistic models.
- Uncertainty analysis cannot be used in hydrological and hydraulic hypothesis testing.
- Uncertainty distributions cannot be understood by policy makers and the public.

Uncertainty can be epistemic uncertainty (originated from imperfection of our knowledge) or variability uncertainty (originated from inherent variability in behavior of natural or human systems).

Uncertainty in modeling......
a. has not been estimated in any water-related study so far.
b. is a maximum of 20%.
c. cannot be understood by policy makers and the public.
d. has a significant impact on the final decision.

I. **Translate the following passage into your mother language. Write your translation in the space provided.**

Fuzzy rule-based systems are data-driven modeling techniques that use fuzzy logic for inference. Fuzzy logic is based on fuzzy set theory in which binary set membership has been extended to include partial membership ranging between 0 and 1. Fuzzy sets have gradual transition between defined sets, allowing direct modeling of uncertainties associated to these concepts. Fuzzy rule-based systems can be created by interviewing human experts, or by processing historical data and thus forming a data-driven model. Fuzzy logic has found multiple successful applications, especially in control theory.

Chaos theory and non-linear dynamics is another data-driven modeling technique that can be used for prediction of time series, if the time series data have enough length and information of the system behavior. The basic idea is to display the system state at time t using a vector in m-dimensional state space. When the original time series has chaotic properties, its equivalent trajectory in phase space has properties that allow it to accurately predict the future values of the independent variable.

Support vector machines (SVM), is a relatively new important data-driven modeling technique. SVM is based on the development of the idea of identifying a line (or a plane or some surface) that separates two classes in classification. It is inspired by V. Vapnik, who initiated the theory of statistical learning in the 1970s.

J. **Copy the technical terms and expressions used in this lesson. Then find your mother language equivalents of those terms and expressions and write them in the spaces provided.**

Technical term	Mother language equivalent
....	....
....	....
....	....
....	....
....	....
....	....
....	....
....	....
....	....
....	....

BIBLIOGRAPHY

Abbott, M.B. (1991). *Hydroinformatics: Information Technology and the Aquatic Environment.* Avebury Technical.

Abrahart, R.J., See, L.M., & Solomatine, D.P. (Eds.). (2008). *Practical Hydroinformatics: Computational Intelligence and Technological Developments in Water Applications* (Vol. 68). Springer Science & Business Media.

Chen, Q., Morales-Chaves, Y., Li, H., & Mynett, A.E. (2006). Hydroinformatics techniques in eco-environmental modelling and management. *Journal of Hydroinformatics* 8(4): 297–316.

De la Fuente, A., Meruane, V., & Meruane, C. (2019). Hydrological early warning system based on a deep learning runoff model coupled with a meteorological forecast. *Water* 11(9): 1808.

Gourbesville, P. (2014). Hydroinformatics and its role in flood management. *Hydrometeorological Hazards: Interfacing Science and Policy*, 137–169.

http://hic2020.org
https://builtin.com/artificial-intelligence
https://en.wikipedia.org/wiki/Agent-based_model
https://en.wikipedia.org/wiki/Artificial_neural_network
https://en.wikipedia.org/wiki/Desertification
https://en.wikipedia.org/wiki/Heuristic
https://en.wikipedia.org/wiki/Hydroinformatics
https://en.wikipedia.org/wiki/Interdisciplinarity
https://en.wikipedia.org/wiki/Machine_learning
https://en.wikipedia.org/wiki/Network_topology
https://en.wikipedia.org/wiki/Synapse
https://www.britannica.com/science/mapping
https://www.expert.ai/blog/machine-learning-definition/
https://www.limswiki.org/index.php/Hydroinformatics
https://www.netapp.com/artificial-intelligence/what-is-artificial-intelligence/
https://www.pcmag.com/encyclopedia/term/integrated-system

https://www.tudelft.nl/en/ceg/about-faculty/departments/hydraulic-engineering

Makropoulos, C., & Savić, D.A. (2019). Urban hydroinformatics: past, present and future. *Water* 11(10): 1959.

Nissani, M. (1995). Fruits, Salads, and Smoothies: A Working definition of Interdisciplinarity. *The Journal of Educational Thought (JET)/Revue de la Pensée Éducative.* 29(2): 121–128. JSTOR 23767672.

Price, R.K., & Solomatine, D.P. (2009). *A brief guide to hydroinformatics.*

See, L., Solomatine, D., Abrahart, R., & Toth, E. (2007). *Hydroinformatics: computational intelligence and technological developments in water science applications. Hydrological Sciences Journal, 52(3), 391–396.*

Vojinovic, Z., & Abbott, M.B. (2017). *Twenty-five years of hydroinformatics.*

GLOSSARY OF TERMS

Artificial intelligence: A wide-ranging branch of computer science concerned with mimicking human intelligence processes through the creation and application of algorithms built into a dynamic computing environment.

Computational hydraulics: An applied science consists of a combination of various disciplines (such as applied mathematics, fluid mechanics, numerical analysis, and computational science) with the aim of simulating various physical processes (involved in seas, estuaries, rivers, channels, lakes) by computers.

Desertification: A type of land degradation in dryland ecosystems caused by natural processes or human activities.

Heuristic: In mathematical optimization and computer science, it is a problem-solving technique that is not guaranteed to be optimal, perfect, or rational, however, is sufficient for reaching an immediate, short-term goal or approximation.

Integrated system: A system that has combined different functions together to work as a unit.

Interdisciplinary: The combination of two or more academic disciplines into one activity.

Mapping: Any prescribed way of assigning to each object in one set a particular object in another (or the same) set.

Machine learning: The study of computer algorithms, as an application of artificial intelligence that provides systems the ability to automatically learn and improve from experience without being explicitly programmed.

Network topology: The arrangement of the elements (links, nodes, etc.) in a communication network.

Synapse: A structure in the nervous system that permits a neuron to pass an electrical or chemical signal to another neuron or to the target effector cell.

5 Hydrogeology

READING FOR COMPREHENSION

Hydrogeology is a branch of geology that studies the distribution and movement of groundwater in the soil and rocks of the Earth's crust. Hydrogeology is an interdisciplinary subject that needs knowledge in several various fields in both the theoretical and the experimental area because it can be complex to determine the chemical, physical, biological, and legal interactions between soil, water, nature, and society.

In the terminology of hydrogeology, hydro means water and geology means the study of the Earth. The terms groundwater hydrology, geohydrology, and ground-water engineering can be used instead of hydrogeology.

AQUIFERS

An aquifer is an underground layer consisting of water-bearing permeable rock, rock fractures, or unconsolidated materials (gravel, sand, or silt). Aquifers produce a feasible quantity of water to be used through a spring or a well. The flow of groundwater from one aquifer to another restricts by aquitards. Aquitards consists of clay or non-porous layers of rock with low hydraulic conductivity. Aquifers are classified in different ways based on their characteristics (Figure 5.1).

SATURATED OR UNSATURATED

The saturated zone, also called the phreatic zone, means all available spaces are filled with water, and the unsaturated zone, also called the vadose zone, means there are still pores of air that contain some water but can be filled with more water.

CONFINED OR UNCONFINED

The confined aquifers are overlain by a confining layer (often made up of clay). The confining layer may protect the aquifer from surface contamination. In an unconfined aquifer, the upper boundary is the water table or phreatic surface, and there is no confining layer between it and the surface (Figure 5.2).

ISOTROPIC OR ANISOTROPIC

In isotropic aquifers, the hydraulic conductivity (K) is equal for flow in all directions, while in anisotropic aquifers it differs, notably in horizontal (K_h) and vertical (K_v) sense.

DOI: 10.1201/9781003293507-5

FIGURE 5.1   Garab spring of Razavi Khorasan province, Iran; (https://www.nabro.ir/garab-spring/).

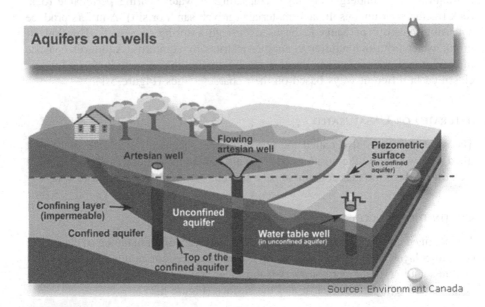

FIGURE 5.2   Schematic image of confined and unconfined aquifers; (https://www.researchgate.net/figure/Adequate-image-for-CONFINED-and-UNCONFINED-AQUIFER_fig1_303738744).

POROUS, KARST, OR FRACTURED

Porous aquifers typically developed in sand and sandstone; therefore, they are also called sandy aquifers. Properties of porous aquifers depend on the depositional sedimentary environment and the sand grains cementation that occurs in a natural process. Karst aquifers typically occur in limestone. Surface water containing natural carbonic acid flows into small fissures in limestone and makes them larger. As a result, more water can enter these fissures and cause the fissures to become larger in a progressive process. Smaller openings store high amounts of water and the larger openings create a conduit system that drains the aquifer to springs. Fractured aquifers are formed during a process in fractured rocks. When a rock with low porosity fractures highly, it can form a good aquifer, if the rock has enough hydraulic conductivity to contribute movement of water (Figure 5.3).

HYDRAULIC HEAD

Hydraulic head, also called piezometric head, is a specific measurement of liquid pressure above a vertical datum that is composed of pressure head (ψ) and elevation head (z). The ψ term can take on negative (suction situation) or positive (saturated situation) values. A pressure transducer can measure this parameter. The z term can be measured relative to an arbitrary datum; therefore, hydraulic head can be any value. The change in hydraulic head per length of flow path, termed head gradient, cause water to move from location of high h to location of low h. Among the relevant terms, hydrograph and drawdown can be mentioned. Hydrograph is a record of hydraulic head during a period at a well, and drawdown is the changes of hydraulic head recorded through a pumping test of a well (Figure 5.4).

FIGURE 5.3 A karst aquifer; (https://www.tehrantimes.com/news/423016/Iran-in-pursuit-of-karst-water-resources).

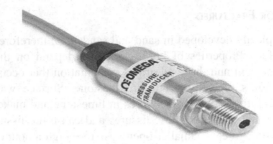

FIGURE 5.4 A pressure transducer; (https://www.elprocus.com/pressure-transducer-working-and-its-applications/).

POROSITY

Porosity (n) is a directly measurable property of aquifer that describes the fraction of void space in the material, where the void may contain, for example, air or water. This parameter is used in hydrogeology, geology, soil science, and building science. The porosity of a porous medium is a fraction between 0 and 1 that is defined by the ratio:

$$n = \frac{V_V}{V_T}$$

where V_V is the volume of void-space and V_T is the total or bulk volume of material, including the solid and void components.

Although porosity has no direct effect on the hydraulic head distribution in an aquifer but it affects groundwater flow velocities through an inversely proportional relationship and subsequently it has a great effect on the dissolved contaminants movement (Figure 5.5).

WATER CONTENT

Water content or moisture content is the fraction of the total rock that is filled with liquid water. Water content is expressed as a ratio, which can range from 0 (completely dry) to 1 (saturated), also it is always less than or equal to porosity.

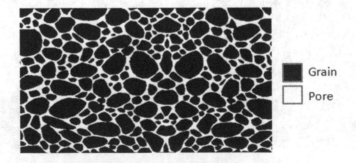

■ Grain
□ Pore

FIGURE 5.5 Microscopic cross section of a porous medium; (https://perminc.com/resources/fundamentals-of-fluid-flow-in-porous-media/chapter-2-the-porous-medium/porosity/).

The unsaturated hydraulic conductivity is a non-linear function of water content. In the vadose zone, the reduction of water content reduces the connected wet pathways through the media, as a result the hydraulic conductivity decreases with lower water content in a strongly non-linear manner. This causes more complexity in the solution of the unsaturated groundwater flow equation, therefore the water content term is very important in vadose zone hydrology.

STORAGE PROPERTIES

In hydrogeology, storage properties are defined as physical properties that characterize the aquifer capacity to release groundwater. These properties are explained as follow:

Specific storage (Ss) is the amount of water that is released from storage by a portion of aquifer, per unit mass or volume of aquifer, per unit change in hydraulic head, while remaining saturated.

Specific yield (Sy), also called the drainable porosity, indicates the amount of water released due to drainage from lowering the water table in an unconfined aquifer that expressed as:

$$S_y = \frac{V_{dw}}{V_T}$$

where V_{dw} the volume of is drained water and V_T is the total rock or material volume.

This ratio is between 0 and 1. The upper boundary of specific yield value is effective porosity because some water will remain in the medium even after drainage due to intermolecular forces; also, specific yield is usually larger than Ss.

Storativity (S), also known as the storage coefficient, is the volume of released water from storage per unit decline in hydraulic head in the aquifer, per unit area of the aquifer. Storativity is a dimensionless quantity that is greater than 0.

These aquifer properties cannot be measured directly but typically determined using some combination of field tests and laboratory tests on aquifer material samples. In addition, recently, remote sensing data obtained from Interferometric synthetic-aperture radar have been used to determine these properties.

CONTAMINANT TRANSPORT PROPERTIES

The contaminants in the aquifers can be natural (arsenic, salinity, etc.) or man-made (petroleum products, nitrate, Chromium or radionuclides, etc.). To understand how dissolved contaminants transport by groundwater, in addition to knowing where the groundwater is flowing; some other aquifer properties that affect contaminants transport should also be considered. These properties are discussed below.

HYDRODYNAMIC DISPERSION

Hydrodynamic dispersivity is an empirical factor that quantifies the deviation of contaminants from the groundwater path. There are longitudinal dispersivity (αL), and transverse dispersivity (αT) that create by macroscopic or microscopic mechanisms. Over long distances in an aquifer, there may be macroscopic in homogeneities that can have areas with higher or lower permeability. This allows some water to move in one direction, and some other in a different direction, so that the contamination spreads quite irregularly. This process is macroscopic mechanism. The microscopic mechanism occurs because water particles may choose any direction (left, right, up, or down) when passing through soil particles; therefore, the water particles and their solute are gradually spread in all directions around the main path. The effect of the macroscopic mechanism on hydrodynamic dispersion is more important than the microscopic mechanism.

Molecular Diffusion

Diffusion is the net movement of anything (e.g. atoms, ions, molecules, energy) that is driven by concentration gradient, from a higher concentration region to a lower concentration region. This fundamental physical phenomenon depends on the random movement of small particles. Diffusion is important for small distances because it is essential for meeting thermodynamic equilibria while it is ineffective for spreading a solute over macroscopic distances because the necessary time to cover a distance by diffusion is proportional to the square of the distance itself. The diffusion phenomenon is quantified by the diffusion coefficient (D) which can often be considered negligible due to its very small amounts, except in cases where the groundwater flow velocities are extremely low (such as clay aquitards).

Retardation by Adsorption

The retardation factor is a very effective factor that causes deviation of the contaminant from the mean groundwater path. The adsorption of the contaminant to the soil holds it back and does not allow it to move forward before the quantity corresponding to the chemical adsorption equilibrium has been adsorbed. This chemical-physical effect changes the contaminant average velocity, so that the contaminant moves much slower than water. The retardation factor depends on the chemical characteristics of the aquifer and the contaminant. As a result of this phenomenon, only more soluble contaminants can cover long distances while less soluble contaminants can move even hundreds or thousands times slower than water.

Pioneering Hydrogeologist

Henry Philibert Gaspard Darcy (1803–1858) was a French engineer who is considered as the founder of the quantitative hydrogeology. In 1855 and 1856, Darcy conducted experiments that studied the movement of fluids through sand columns. These experiments led to the determination of Darcy's law that was initially developed to describe flow through sands but now it has been widely used and generalized to a variety of situations.

Oscar Edward Meinzer (1876–1948) was an American scientist who was known as the "father of modern groundwater hydrology" due to his studies and achievements.

The Meinzer achievements that can be mentioned are including:

- Standardizing the key terms in this field,
- Determining the principles regarding occurrence, movement, and discharge,
- Proving that the water flow follows Darcy's law,
- Suggesting the use of geophysical methods and recorders on wells,
- Proposing the utilize of pumping tests to collect quantitative information on the properties of aquifers,
- Highlighting the importance of considering the geochemistry of water and the high salinity effect in aquifers.

GOVERNING EQUATIONS

Darcy's law

Darcy's law is the fundamental equation to understand the fluids movement in the Earth's crust. Darcy's law states, "The amount of groundwater discharging through a column of soil is directly proportional to the hydraulic conductivity, cross-sectional area of flow, and the hydraulic head difference and inversely proportional to the length of the column."

The mathematical form of Darcy's law is as follows:

$$Q = -K\,(A\Delta h/l)$$

where Q is volumetric flow rate (m³/s), K is hydraulic conductivity (m/s), A is cross-sectional area of flow (m²), Δh is hydraulic head difference over the path l (m), and l = flow path length (m).

The minus sign on the right side of the equation indicates that the fluid flows from points with high potential energy to points with low potential energy therefore hydraulic head always reduces in the direction of flow.

There are limits to Darcy's law application. Darcy's law is valid for laminar flow; however, most of groundwater flow cases are covered by Darcy's law. The nature of flow is quantified by the Reynolds number, a dimensionless ratio, expressed as:

$$Re = \rho v d/\mu$$

where ρ is the fluid density (kg/m³), v is the flow velocity (m/s), d is the diameter of pipe (m), and μ is the fluid viscosity (kg/m/s).

Typically, flow regimes with Reynolds numbers less than one are clearly laminar. Experiments indicate that flow with a Reynolds numbers up to 10 may still be valid to apply Darcy's law, as in the case of groundwater flow.

GROUNDWATER FLOW EQUATION

The groundwater flow equation is a mathematical relationship in differential equation form, which is used in hydrogeology to describe the movement of groundwater in the porous media. The groundwater flow is analyzed by coupling transport law (Darcy's law) and **mass** conservation law (continuity equation) with the specified initial and boundary conditions. The Laplace equation and diffusion equation are used to describe the steady-state flow and the transient flow, respectively.

There are two broad categories to achieving solutions to a mathematical model of groundwater flow: analytical or numerical solution. Analytic methods typically use the structure of mathematics to give a quick answer based on a few basic parameters, but the required derivation for all (involving non-standard coordinates, conformal mapping, etc.) may be very complex, except the simplest domain geometries. One of the most commonly analytic solutions to the groundwater flow equation is the Theis equation that is very simple and still very useful. This equation can be used to predict the transient evolution of head due to the pumping wells. Numerical methods that are very extensive have a long history. In the 1920s, Lewis Fry Richardson, an English scientist, developed some finite difference schemes still in use today, but today these methods have become very important because of the use of personal computers instead of hand calculations. Various numerical solution methods are including general properties of gridded methods, application of finite difference models, application of finite element models, application of finite volume models, etc.

EXERCISES

A. **Read each statement and decide whether it is true or false. Write "T" for true and "F" for false statements.**

TF1. The term groundwater hydrology can be used instead of hydrogeology.

TF2. In the phreatic zone, there are still pores of air that contain some water but can be filled with more water.

TF3. The ψ term is positive in suction situation and negative in saturated situation.

TF4. Porosity is a fraction between 0 and 1 that has no direct effect on the hydraulic head distribution in an aquifer.

TF5. Specific storage, specific yield, and storativity can be determined by remote sensing data obtained from Interferometric synthetic-aperture radar.

B. **Circle a, b, c, or d which best completes the following items.**

1. .... is a specific measurement of liquid pressure above a vertical datum
   a. Moisture content
   b. Head gradient
   c. Hydraulic head
   d. Pressure head

2. .... is the fraction of the total rock that is filled with liquid water.
   a. Saturated zone
   b. Moisture content
   c. Porosity
   d. Specific storage

3. The nature of flow is quantified by the ....
   a. retardation factor.
   b. Darcy's law.
   c. diffusion coefficient.
   d. Reynolds number.
4. ...…............ is the net movement of anything that is driven by concentration gradient.
   a. Diffusion
   b. Head gradient
   c. Hydrodynamic dispersivity
   d. Drawdown
5. .... is an analytic solution to the groundwater flow equation that is very simple and still very useful.
   a. Continuity equation
   b. Theis equation
   c. Laplace equation
   d. Diffusion equation

C. **Match the sentence halves in Column I with their appropriate halves in Column II. Insert the letters a, b, c ... in the parentheses provided. There are more sentence halves in Column II than required.**

Column I	Column II
1. Spring	() **a.** indicates the amount of water released due to drainage from lowering the water table in an unconfined aquifer.
2. Aquitards	() **b.** records the hydraulic head during a period at a well.
3. Pressure transducer	() **c.** a deep hole sunk into the ground that was created to obtain liquid resources.
4. Specific yield	() **d.** the volume of released water from storage per unit decline in hydraulic head in the aquifer, per unit area of the aquifer.
5. Well	() **e.** measures the pressure of a fluid through converting it into an electrical signal.
	() **f.** a place where water flows from an aquifer to the land surface.
	() **g.** used in hydrogeology to describe the movement of groundwater in the porous media.
	() **h.** restricts the flow of groundwater from one aquifer to another.

D. **Cross out the word or words that make each statement false, and write the word or words that make each statement true in the blank.**
   1. The unsaturated hydraulic conductivity is a function of water content in a linear manner.

      ....

2. As a result of the retardation by adsorption, only less soluble contaminants can cover long distances.

   ....

3. Darcy's law states, "The amount of groundwater discharging through a column of soil is inversely proportional to the hydraulic conductivity, cross-sectional area of flow, and the hydraulic head difference and directly proportional to the length of the column."

   ....

4. Henry Darcy was known as the father of modern groundwater hydrology and Oscar Meinzer considered as the founder of the quantitative hydrogeology.

   ....

E. **Give answers to the following questions.**
   1. What is the process of formation of karst aquifers?
   2. What is the macroscopic and microscopic mechanism in hydrodynamic dispersion?
   3. What are the achievements of Meinzer?
   4. What are the limitations of using Darcy Law?

F. **For each word on the left, there are three meanings provided. Put a check mark (√) next to the choice which has the closest meaning to the word given.**

1. Transient flow	steady flow	unsteady flow	laminar flow
2. Porous aquifers	limestone aquifers	fractured aquifers	sandy aquifers
3. Piezometric head	pressure head	elevation head	hydraulic head systems
4. Mass conservation law	continuity equation	Laplace equation	groundwater flow equation
5. Phreatic surface	confining layer	vadose zone	water table

G. **Fill in the blanks with the appropriate words from the following list.**
   pumping test   Laplace equation   molecular diffusion   storativity
   diffusion equation   spring   fractured   Reynolds number

   Hydrogeology is an interdisciplinary field that studies the distribution and movement of groundwater. Aquifers are underground layers which their water can be obtained through a ....or a well. Well characteristics can estimate through a .... The aquifer capacity to release groundwater depends on specific storage, specific yield and ....There are three types of aquifers based on their formation; porous, karst, or .... Contaminants are a concern about groundwater that must be managed. For this purpose, it is necessary to know the effective aquifer properties in the movement of contaminants, which are hydrodynamic dispersion ... and retardation by adsorption. Darcy's law is the basic equation for analyzing the movement of fluids in the Earth's crust, the validity range of which is determined by the .... The groundwater flow equation is a differential equation, which is used to describe the groundwater movement. Specifically, the steady-state flow is described by the ....and the transient flow by the ....

H. **Read this passage and then circle a, b, c, or d which best completes the following items.**
Increasing population causes water stress in various sources, including groundwater. This situation requires the implementation of new water policies in some urban areas, which move cities proactively to conserve groundwater. This has known as proactive land-use management. For example in Brazil case, overpopulation caused municipally provided water to run low. Because of water scarcity, people began to create wells that were normally within the municipal water system. This solution was possible only for a class of society that was in high socioeconomic standing, but many of the impoverished population remained without access to water. At this point, the municipality came in and implemented new policies to help those who could not afford to drill wells of their own. Drilling wells by the municipality enables them to plan for the future sustainability of groundwater in the area, taking into account growing populations.

Some Brazilians could not drill their wells of their own because....
a. they did not have the technical ability for drilling a well.
b. the municipality only allowed drilling wells for money.
c. they could not afford to drill a well.
d. the water laws did not allow them.

I. **Translate the following passage into your mother language. Write your translation in the space provided.**
Water well is a mechanism to obtain groundwater by drilling or digging and delivering water to the surface using pump or hand. Some of the earliest evidence of water wells has been found in China. Old Chinese documents and archeological evidence indicate that the prehistoric and ancient Chinese had the aptitude and skill of digging deep-water wells from 6,000 to 7,000 years ago. They discovered and made extensive use of deep drilled groundwater for drinking.
    Wells are traditionally been sunk by hand digging, which is both inexpensive and low-tech. The structure can be lined with brick or

stone as the excavation proceeds. In addition, there is a more modern method for drilling a well called caisson that uses pre-cast reinforced concrete well rings that are lowered into the hole.

There are three main types of wells, shallow, deep, and artesian. Shallow wells access unconfined aquifers, and are generally shallow (less than 15 m deep) and small in diameter (typically less than 15 cm). Deep wells tap into confined aquifers, and are always drilled using machine. All deep wells bring water to the surface by mechanical pumps. In artesian wells, water flows naturally without a pump or some other mechanical device because the top of the well is being located below the water table.

J. **Copy the technical terms and expressions used in this lesson. Then find your mother language equivalents of those terms and expressions and write them in the spaces provided.**

Technical term	Mother language equivalent
....	....
....	....
....	....
....	....
....	....
....	....
....	....
....	....
....	....
....	....

BIBLIOGRAPHY

Ahmed, S., Jayakumar, R., & Salih, A. (Eds.). (2008). *Groundwater Dynamics in Hard Rock Aquifers: Sustainable Management and Optimal Monitoring Network Design*. India: Springer Science & Business Media.

Béjar-Pizarro, M., Ezquerro, P., Herrera, G., Tomás, R., Guardiola-Albert, C., Hernández, J.M.R., ... & Martínez, R. (2017). Mapping groundwater level and aquifer storage variations from InSAR measurements in the Madrid aquifer, Central Spain. *Journal of Hydrology*, 547, 678–689.

Chanson, H. (2004). Hydraulics of Open Channel Flow: An Introduction. *Butterworth–Heinemann*, ISBN 978-0750659789, 650 pages. See p. 22.

Dreybrodt, W. (1988). Processes in karst systems: physics, chemistry, and geology. *Springer Series in Physical Environment*. 4. Berlin: Springer. pp. 2–3. doi:10.1007/978-3-642-83352-6. ISBN 978-3-642-83354-0.

Foster, S.D., Hirata, R., Howard, K.W.F. (2010). Groundwater use in developing cities: Policy issues arising from current trends. *Hydrogeology Journal* 19(2): 271–274. doi:10.1007/s10040-010-0681-2.

Harter, T. ANR Publication 8086. Water Well Design and Construction http://groundwater.ucdavis.edu/files/156563.pdf.

https://bae.okstate.edu/faculty-sites/Darcy/1pagebio.htm

http://www.wateringmalawi.org/Watering_Malawi/Resources_files/Boreholewells.pdf

https://www2.gov.bc.ca

https://en.wikipedia.org/wiki/Aquifer

https://en.wikipedia.org/wiki/Darcy[,]s_law

https://en.wikipedia.org/wiki/Diffusion

https://en.wikipedia.org/wiki/Groundwater_flow_equation

httpoı//on.wikipedia.org/wiki/Henry_Darcy

https://en.wikipedia.org/wiki/Hydraulic_head

https://en.wikipedia.org/wiki/Hydrogeology

https://en.wikipedia.org/wiki/Lewis_Fry_Richardson

https://en.wikipedia.org/wiki/Oscar_Edward_Meinzer

https://en.wikipedia.org/wiki/Phreatic

https://en.wikipedia.org/wiki/Porosity

https://en.wikipedia.org/wiki/Specific_storage

https://en.wikipedia.org/wiki/Spring_(hydrology)

https://en.wikipedia.org/wiki/Water_content

https://en.wikipedia.org/wiki/Well

https://en.wikipedia.org/wiki/Well

http://go.galegroup.com/ps/i.do?id=GALE%7CCX2830902895&v=2.1&u=nclivensu&it=r&p=GVRL&sw=w&asid=88753af7557df17de94c1979354d8c74

https://perminc.com/resources/fundamentals-of-fluid-flow-in-porous-media/chapter-2-the-porous-medium/porosity/

http://www.aqtesolv.com

https://www.biologyonline.com

https://www.britannica.com

https://www.dictionary.com/browse/well

https://www.elprocus.com/pressure-transducer-working-and-its-applications/

https://www.ksb.com

https://www.nabro.ir/garab-spring/

https://www.omega.com

https://www.researchgate.net/figure/Adequate-image-for-CONFINED-and-UNCONFINED-AQUIFER_fig1_303738744

https://www.surecontrols.com
https://www.tehrantimes.com/news/423016/Iran-in-pursuit-of-karst-water-resources
https://www.toppr.com
https://www.usgs.gov
"Introduction to Ground Water Extraction Technologies: Borehole, Shallow Well, and Tube Well".
Kuhn, O. (2004-06-30). Ancient Chinese Drilling. *Canadian Society of Exploration Geophysicists* 29(6).
Mulley, R. (2004). *Flow of Industrial Fluids: Theory and Equations*. USA: CRC Press.
Oklahoma State. Henry Darcy and His Law September 3, 2003.
Schetz, J.A., Fuhs, A.E. (ed.). (1999). *Fundamentals of Fluid Mechanics*. USA: John Wiley & Sons.
Tomás, R., Herrera, G., Delgado, J., Lopez-Sanchez, J.M., Mallorquí, J.J., & Mulas, J. (2010). A ground subsidence study based on DInSAR data: Calibration of soil parameters and subsidence prediction in Murcia City (Spain). *Engineering Geology*, 111(1–4), 19–30.

GLOSSARY OF TERMS

Concentration gradient: The gradual change in the solutes concentration in a solution as a function of distance through a solution.

Phreatic surface: Where the pore water pressure is under atmospheric conditions that typically coincides with the water table.

Pressure transducer: A device that measures the pressure of a fluid (the force that the fluid is exerting on surfaces in contact with it) by converting it into an electrical signal.

Pumping test: A test to estimate hydraulic properties of an aquifer system is as follows: Pumping groundwater from a well, and measuring the change in water level in the pumping well and any adjacent wells or surface water bodies during and after pumping.

Spring: A place where water flows from an aquifer to the land surface and emerges.

Steady-state flow: A steady flow is defined as that in which the various parameters (velocity, pressure, and cross-section) at any point do not change with time.

Thermodynamic equilibrium: A state of a thermodynamic system in which the system is in mechanical, chemical, and thermal equilibrium, and there is no tendency for spontaneous change.

Transient flow: Transient or unsteady flow is a fluid dynamics condition in which flow parameters (i.e. velocity and pressure) change over time due to changes in system status.

Well: A deep hole or shaft sunk into the ground that is created by digging, driving, or drilling to obtain liquid resources, usually water.

6 Hydrology science

READING FOR COMPREHENSION

Hydrology is the study of the movement, distribution, and quality of water throughout the Earth, including the hydrologic cycle, water resources, and environmental watershed sustainability. A practitioner of hydrology is a hydrologist, working within the fields of either earth or environmental science, physical geography, geology, or civil and environmental engineering.

Domains of hydrology include hydrometeorology, surface hydrology, hydrogeology, drainage basin management, and water quality, where water plays a central role. Oceanography and meteorology are not included because water is only one of many important aspects.

Hydrological research can inform environmental engineering, policy, and planning.

The term *hydrology* is from Greek: ὕδωρ, *hydōr*, "water"; and λόγος, *logos*, "study" (Figure 6.1).

HISTORY OF HYDROLOGY

Hydrology has been a subject of investigation and engineering for millennia. For example, about 4000 BC the Nile was dammed to improve agricultural productivity of previously barren lands. Mesopotamian towns were protected from flooding with high earthen walls. Aqueducts were built by the Greeks and Ancient Romans, while the History of China shows they built irrigation and flood control works. The ancient Sinhalese used hydrology to build complex irrigation Works in Sri Lanka, also known for invention of the Valve Pit which allowed construction of large reservoirs, anicuts, and canals that still function.

Marcus Vitruvius, in the 1st century B.C., described a philosophical theory of the hydrologic cycle, in which precipitation falling in the mountains infiltrated the Earth's surface and led to streams and springs in the lowlands. With adoption of a more scientific approach, Leonardo da Vinci and Bernard Palissy independently reached an accurate representation of the hydrologic cycle. It was not until the 17th century that hydrologic variables began to be quantified.

Pioneers of the modern science of hydrology include Pierre Perrault, Edme Mariotte, and Edmund Halley. By measuring rainfall, runoff, and drainage area, Perrault showed that rainfall was sufficient to account for flow of the Seine. Marriotte combined velocity and river cross-section measurements to obtain discharge, again in the Seine. Halley showed that the evaporation from the Mediterranean Sea was sufficient to account for the outflow of rivers flowing into the sea.

Advances in the 18th century included the Bernoulli piezometer and Bernoulli's equation, by Daniel Bernoulli, the Pitot tube. The 19th century saw development in

DOI: 10.1201/9781003293507-6

FIGURE 6.1 Water covers 70% of the Earth's surface, Marmara Sea, Turkey (By Mohammad Albaji).

groundwater hydrology, including Darcy's law, the Dupuit-Thiem well formula, and Hagen-Poiseuille's capillary flow equation.

Rational analyses began to replace empiricism in the 20th century, while governmental agencies began their own hydrological research programs. Of particular importance were Leroy Sherman's unit hydrograph, the infiltration theory of Robert E. Horton, and C.V. Theis's Aquifer test/equation describing well hydraulics.

Since the 1950s, hydrology has been approached with a more theoretical basis than in the past, facilitated by advances in the physical understanding of hydrological processes and by the advent of computers and especially Geographic Information Systems (GIS).

HYDROLOGIC CYCLE

The central theme of hydrology is that water circulates throughout the Earth through different pathways and at different rates. The most vivid image of this is in the evaporation of water from the ocean, which forms clouds. These clouds drift over the land and produce rain. The rainwater flows into lakes, rivers, or aquifers. The water in lakes, rivers, and aquifers then either evaporates back to the atmosphere or eventually flows back to the ocean, completing a cycle. Water changes its state of being several times throughout this cycle (Figure 6.2).

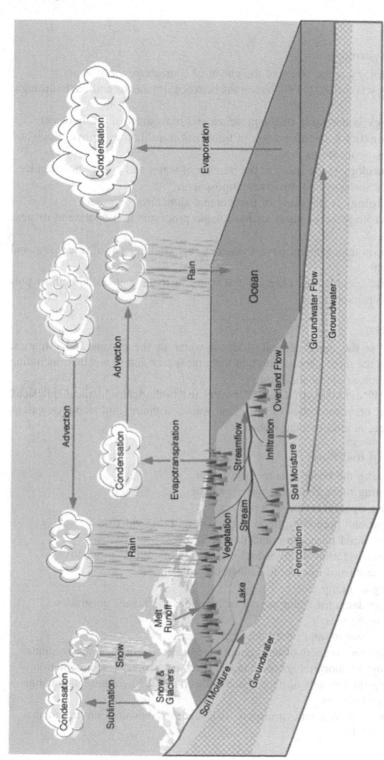

FIGURE 6.2 Hydrologic cycle; (Physical geography.net).

OVERVIEW

Branches of hydrology

Chemical hydrology is the study of the chemical characteristics of water.

Ecohydrology is the study of interactions between organisms and the hydrologic cycle.

Hydrogeology is the study of the presence and movement of groundwater.

Hydro informatics is the adaptation of information technology to hydrology and water resources applications.

Hydrometeorology is the study of the transfer of water and energy between land and water body surfaces and the lower atmosphere.

Isotope hydrology is the study of the isotopic signatures of water.

Surface hydrology is the study of hydrologic processes that operate at or near Earth's surface.

Drainage basin management covers water storage, in the form of reservoirs, and flood protection.

Water quality includes the chemistry of water in rivers and lakes, both of pollutants and natural solutes.

Related topics

Oceanography is the more general study of water in the oceans and estuaries. Meteorology is the more general study of the atmosphere and of weather, including precipitation such as snow and rainfall.

Limnology is the study of lakes. It covers the biological, chemical, physical, geological, and other attributes of all inland waters (running and standing waters, fresh and saline, natural or man-made).

Applications of hydrology

- Determining the water balance of a region.
- Determining the agricultural water balance.
- Designing riparian restoration projects.
- Mitigating and predicting flood, landslide, and drought risk.
- Real-time flood forecasting and flood warning.
- Designing irrigation schemes and managing agricultural productivity.
- Part of the hazard module in catastrophe modeling.
- Providing drinking water.
- Designing dams for water supply or hydroelectric power generation.
- Designing bridges.
- Designing sewers and urban drainage systems.
- Analyzing the impacts of antecedent moisture on sanitary sewer systems.
- Predicting geomorphological changes, such as erosion or sedimentation.
- Assessing the impacts of natural and anthropogenic environmental change on water resources.
- Assessing contaminant transport risk and establishing environmental policy guidelines.

Hydrologic measurements

Measurement is fundamental for assessing water resources and understanding the processes involved in the hydrologic cycle. Because the hydrologic cycle is so diverse, hydrologic measurement methods span many disciplines: including soils, oceanography, atmospheric science, geology, geophysics, and limnology, to name a few. Here, hydrologic measurement methods are organized by hydrologic sub-disciplines. Each of these sub-disciplines is addressed briefly with a practical discussion of the methods used to date and a bibliography of background information (Figure 6.3).

Quantifying groundwater flow and transport

- Aquifer characterization
  - Flow direction
    - Piezometer – groundwater pressure and, by inference, groundwater depth (see: aquifer test)
    - Conductivity, storativity, transmissivity
    - Geophysical methods
- Vadose zone characterization
  - Infiltration
    - Infiltrometer – infiltration
  - Soil moisture
    - Capacitance probe-soil moisture
    - Time-domain reflectometer – soil moisture
    - Tensiometer – soil moisture
    - Solute sampling
    - Geophysical methods

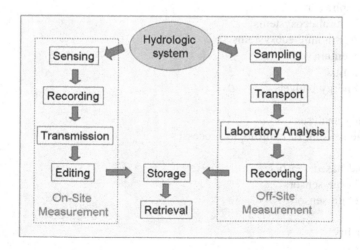

FIGURE 6.3 Hydrologic measurement and the development of a hydrologic observations database; (http://www.crwr.utexas.edu/cuahsi/symposium05/Introduction/HISmission_0305.htm).

Quantifying surface water flow and transport

- Direct and indirect discharge measurements
  - Stream gauge – stream flow (see: discharge (hydrology))
  - Tracer techniques
  - Chemical transport
  - Sediment transport and erosion
  - Stream-aquifer exchange

Quantifying exchanges at the land-atmosphere boundary

- Precipitation
  - Bulk rain events
    - Disdrometer – precipitation characteristics
    - Radar – cloud properties, rain rate estimation, hail, and snow detection
    - Rain gauge – rain and snowfall
    - Satellite – rainy area identification, rain rate estimation, land-cover/land-use, soil moisture
    - Sling psychrometer – humidity
  - Snow, hail, and ice
  - Dew, mist, and fog
- Evaporation
  - from water surfaces
  - Evaporation – Symon's evaporation pan
  - from plant surfaces
  - through the boundary layer
- Transpiration
  - Natural ecosystems
  - Agronomic ecosystems
- Momentum
- Heat flux
  - Energy budgets

Uncertainty analyses
Remote sensing of hydrologic processes

- Land-based sensors
- Airborne sensors
- Satellite sensors

Water quality

- Sample collection
- In-situ methods
- Physical measurements (includes sediment concentration)

- Collection of samples to quantify organic compounds
- Collection of samples to quantify inorganic compounds
- Analysis of aqueous organic compounds
- Analysis of aqueous inorganic compounds
- Microbiological sampling and analysis

Integrating measurement and modeling

- Budget analyses
- Parameter estimation
- Scaling in time and space
- Data assimilation
- Quality control of data – see for example double mass analysis

HYDROLOGIC PREDICTION

Observations of hydrologic processes are used to make predictions of the future behavior of hydrologic systems (water flow, water quality). One of the major current concerns in hydrologic research is "Prediction in Ungauged Basins" (PUB), i.e. in basins where no or only very few data exist.

STATISTICAL HYDROLOGY

By analyzing the statistical properties of hydrologic records, such as rainfall or river flow, hydrologists can estimate future hydrologic phenomena, assuming the characteristics of the processes remain unchanged.

These estimates are important for engineers and economists so that proper risk analysis can be performed to influence investment decisions in future infrastructure and to determine the yield reliability characteristics of water supply systems. Statistical information is utilized to formulate operating rules for large dams forming part of systems that include agricultural, industrial, and residential demands.

HYDROLOGIC MODELING

Hydrologic models are simplified, conceptual representations of a part of the hydrologic cycle. They are primarily used for hydrologic prediction and for understanding hydrologic processes. Two major types of hydrologic models can be distinguished:

- Models based on data. These models are black box systems, using mathematical and statistical concepts to link a certain input (for instance rainfall) to the model output (for instance runoff). Commonly used techniques are regression, transfer functions, and system identification. The simplest of these models may be linear models, but it is common to deploy nonlinear components to represent some general aspects of a catchment's response without going deeply into the real physical processes involved. An example of such an aspect is the well-known behavior that a catchment will respond much more quickly and strongly when it is already wet than when it is dry.

- Models based on process descriptions. These models try to represent the physical processes observed in the real world. Typically, such models contain representations of surface runoff, subsurface flow, evapotranspiration, and channel flow, but they can be far more complicated. These models are known as deterministic hydrology models. Deterministic hydrology models can be subdivided into single-event models and continuous simulation models.

Recent research in hydrologic modeling tries to have a more global approach to the understanding of the behavior of hydrologic systems to make better predictions and to face the major challenges in water resources management.

HYDROLOGIC TRANSPORT

Water movement is a significant means by which other materials, such as soil or pollutants, are transported from place to place. Initial input to receiving waters may arise from a point source discharge or a line source or area source such as surface runoff. Since the 1960s rather complex mathematical models have been developed, facilitated by the availability of high-speed computers. The most common pollutant classes analyzed are nutrients, pesticides, total dissolved solids, and sediment.

EXERCISES

A. **Read each statement and decide whether it is true or false. Write "T" for true and "F" for false statements.**

TF1. Domains of hydrology include oceanography and meteorology.

TF2. The research of hydrology can inform environmental engineering, policy and planning.

TF3. The central theme of hydrology is that water circulates throughout the Earth.

TF4. Mesopotamian cities were protected from flooding with high earthen dams.

TF5. Ecohydrology is the study of interactions between ecology and the hydrologic cycle.

B. **Circle a, b, c, or d which best completes the following items.**

1. Hydrogeology is the study of the .... of groundwater.
    a. rising
    b. presence and movement
    c. quality and quantity
    d. movement

2. .... is the adaptation of information technology to hydrology and water resources applications.
   a. Hydroinformatics
   b. Surface hydrology
   c. Hydrometeorology
   d. Hydrogeology

3. .... is the study of the transfer of water and energy between land and water body surfaces and the lower atmosphere.
   a. Hydrogeology
   b. Hydroinformatics
   c. Hydrometeorology
   d. Statistical hydrology

4. Limnology is the study of ....
   a. seas.
   b. rivers.
   c. oceans.
   d. lakes.

5. .... is the more general study of water in the estuaries.
   a. Meteorology
   b. Oceanography
   c. Hydrogeology
   d. Limnology

C. **Match the sentence halves in Column I with their appropriate halves in Column II. Insert the letters a, b, c ... in the parentheses provided. There are more sentence halves in Column II than required.**

Column I	Column II
1. Hydrology is the study of the	( ) **a.** hydrologist, working within the fields of either earth or environmental science.
2. The Nile was dammed to	( ) **b.** to make predictions of the future behavior of hydrologic systems.
3. A practitioner of hydrology is a	( ) **c.** movement, distribution, and quality of water throughout the Earth.
4. Observations of hydrologic processes are used	( ) **d.** including the hydrologic cycle, water resources, and environmental watershed sustainability
5. Hydrologic models	( ) **e.** improve agricultural productivity of previously agricultural lands.
	( ) **f.** are organized by hydrologic sub-disciplines.
	( ) **g.** are simplified, conceptual representations of a part of the hydrologic cycle.
	( ) **h.** improve agricultural productivity of previously barren lands.

D. **Cross out the word or words that make each statement false, and write the word or words that make each statement true in the blank.**
   1. The term hydraulic is from Greek: ὕδωρ, *hydōr*, "water"; and λόγος, *logos*, "study".

      ....
   2. Hydrogeology is the study of hydrologic processes that operate at or near Earth's surface.

      ....
   3. Hydrologic models are black box systems, using physical and chemical concepts to link a certain input to the model output.

      ....
   4. Observation is fundamental for assessing water resources and understanding the processes involved in the hydrologic cycle.

      ....

E. **Give answers to the following questions.**
   1. What are the applications of hydrology?
   2. What are the branches of hydrology?
   3. What are the hydrologic measurements?

F. **For each word on the left, there are three meanings provided. Put a check mark (√) next to the choice which has the closest meaning to the word given.**

1. Hydrologic prediction	hydrologic forecasting	hydrologic measurement	hydrologic processes
2. Hydrology	water research	water study	water recourse
3. Infiltration	permeability	conductivity	consolidation
4. Meteorology	study of the lithosphere	study of the atmosphere	study of the mesosphere
5. Isotope hydrology	isotopic relation of water	isotopic measurement of water	isotopic signatures of water

G. **Fill in the blanks with the appropriate words from the following list.**
   prediction   hydrologists   travel   summarizing   data
   performing analyses   pollution   persons   quality   flooding

   .... apply scientific knowledge and mathematical principles to solve water-related problems in society: problems of quantity, ... and availability. They may be concerned with finding water supplies for cities or irrigated farms, or controlling river .... or soil erosion. Or, they may work in environmental protection: preventing or cleaning up ....

or locating sites for safe disposal of hazardous wastes. ....
trained in hydrology may have a wide variety of job titles. Scientists
and engineers in hydrology may be involved in both field
investigations and office work. In the field, they may collect basic
...............; oversee testing of water quality, direct field crews and
work with equipment. Many jobs require ...., some
abroad. A hydrologist may spend considerable time doing fieldwork
in remote and rugged terrain. In the office, hydrologists do many
things such as interpreting hydrologic data and .... for
determining possible water supplies. Much of their work relies on
computers for organizing, ... and analyzing masses of
data, and for modeling studies such as the .... of flooding
and the consequences of reservoir releases or the effect of leaking
underground oil storage tanks.

H. **Read this passage and then circle a, b, c, or d which best
completes the following items.**
Water is one of our most important natural resources. Without it,
there would be no life on earth. The supply of water available for
our use is limited by nature. Although there is plenty of water on
earth, it is not always in the right place, at the right time and of the
right quality. Adding to the problem is the increasing evidence that
chemical wastes improperly discarded yesterday are showing up in
our water supplies today. Hydrology has evolved as a science in
response to the need to understand the complex water systems of the
Earth and help solve water problems. Hydrologists play a vital role
in finding solutions to water problems, and interesting and chal-
lenging careers are available to those who choose to study hydrology.

The main purpose of this passage is ....
a. introduction of hydrologists.
b. introduction of hydrology (Hydrology is the study of water).
c. chemical wastes and contaminants.
d. conservation of water.

I. **Translate the following passage into your mother language.
Write your translation in the space provided.**
Most cities meet their needs for water by withdrawing it from the
nearest river, lake or reservoir. Hydrologists help cities by collecting and analyzing the data needed to predict how much water is
available from local supplies and whether it will be sufficient to
meet the city's projected future needs. To do this, hydrologists
study records of rainfall, snowpack depths, and river flows that are
collected and compiled by hydrologists in various government
agencies. They inventory the extent river flow already is being used
by others.

Managing reservoirs can be quite complex because they generally serve many purposes. Reservoirs increase the reliability of local water supplies. Hydrologists use topographic maps and aerial photographs to determine where the reservoir shorelines will be and to calculate reservoir depths and storage capacity. This work ensures that, even at maximum capacity, no highways, railroads, or homes would be flooded.

Deciding how much water to release and how much to store depends upon the time of year, flow predictions for the next several months, and the needs of irrigators and cities as well as downstream water-users that rely on the reservoir. If the reservoir also is used for recreation or for generation of hydroelectric power, those requirements must be considered. Decisions must be coordinated with other reservoir managers along the river. Hydrologists collect the necessary information, enter it into a computer, and run computer models to predict the results under various operating strategies. On the basis of these studies, reservoir managers can make the best decision for those involved.

The availability of surface water for swimming, drinking, industrial, or other uses sometimes is restricted because of pollution. Pollution can be merely an unsightly and inconvenient nuisance, or it can be an invisible, but deadly, threat to the health of people, plants, and animals.

J. **Copy the technical terms and expressions used in this lesson. Then find your mother language equivalents of those terms and expressions and write them in the spaces provided.**

Technical term	Mother language equivalent
....	....
....	....
....	....
....	....
....	....
....	....
....	....
....	....
....	....
....	....

BIBLIOGRAPHY

Encyclopedia of Hydrological Sciences. ISBN 0-471-49103-9.
Handbook of Hydrology. ISBN 0-07-039732-5.
https://en.wikipedia.org/wiki/Hydrology
http://water.usgs.gov/edu/hydrology.html
http://www.crwr.utexas.edu/cuahsi/symposium05/Introduction/HISmission_0305.htm
http://www.exponent.com/hydrology_and_hydraulics
Hydrologic Analysis and Design. McCuen, Third Edition, 2005. ISBN 0-13-142424-6.
Hydrological Processes, ISSN: 1099-1085 (electronic) 0885-6087 (paper), John Wiley & Sons.
Hydrology Research (formerly Nordic Hydrology), ISSN: 0029-1277, IWA Publishing.
Introduction to Hydrology, 4e. Viessman and Lewis, (1996). ISBN 0-673-99337-X.
Introduction to Physical Hydrology, Martin Hendriks(2010). ISBN 9780199296842.
Journal of Hydroinformatics, ISSN: 1464-7141, IWA Publishing.
Journal of Hydrologic Engineering, ISSN: 0733-9496, ASCE Publication.
Physical geography.net

GLOSSARY OF TERMS

Chemical hydrology: Chemical hydrology is the study of the chemical characteristics of water.

Ecohydrology: Ecohydrology is the study of interactions between organisms and the hydrologic cycle.

Hydrogeology: Hydrogeology is the study of the presence and movement of groundwater.

Hydroinformatics: Hydroinformaticsis the adaptation of information technology to hydrology and water resources applications.

Hydrology: Hydrology is the study of the movement, distribution, and quality of water throughout the Earth, including the hydrologic cycle, water resources, and environmental watershed sustainability.

Hydrometeorology: Hydrometeorology is the study of the transfer of water and energy between land and water body surfaces and the lower atmosphere.

Isotope hydrology: Isotope hydrology is the study of the isotopic signatures of water.

Limnology: Limnology is the study of lakes. It covers the biological, chemical, physical, geological, and other attributes of all inland waters.

Meteorology: Meteorology is the more general study of the atmosphere and of weather, including precipitation such as snow and rainfall.

Oceanography: Oceanography is the more general study of water in the oceans and estuaries.

Surface hydrology: Surface hydrology is the study of hydrologic processes that operate at or near Earth's surface.

7 Water resources science

READING FOR COMPREHENSION

Water resources are sources of water that are useful or potentially useful to humans. Uses of water include agricultural, industrial, household, recreational, and environmental activities. Virtually all of these human uses require freshwater.

Ninety-seven percent of water on the Earth is saltwater, and only 3% is freshwater of which slightly over two-thirds is frozen in glaciers and polar ice caps. The remaining unfrozen freshwater is mainly found as groundwater, with only a small fraction present above ground or in the air.

Freshwater is a renewable resource, yet the world's supply of clean, freshwater is steadily decreasing. Water demand already exceeds supply in many parts of the world and as the world population continues to rise, so too does the water demand. Awareness of the global importance of preserving water for ecosystem services has only recently emerged as, during the 20th century, more than half the world's wetlands have been lost along with their valuable environmental services. Biodiversity-rich freshwater ecosystems are currently declining faster than marine or land ecosystems. The framework for allocating water resources to water users (where such a framework exists) is known as water rights (Figure 7.1).

SOURCES OF FRESHWATER

Surface water

Surface water is water in a river, lake, or freshwater wetland. Surface water is naturally replenished by precipitation and naturally lost through discharge to the oceans, evaporation, evapotranspiration, and sub-surface seepage (Figure 7.2).

Although the only natural input to any surface water system is precipitation within its watershed, the total quantity of water in that system at any given time is also dependent on many other factors. These factors include storage capacity in lakes, wetlands, and artificial reservoirs, the permeability of the soil beneath these storage bodies, the runoff characteristics of the land in the watershed, the timing of the precipitation, and local evaporation rates. All of these factors also affect the proportions of water lost.

Human activities can have a large and sometimes devastating impact on these factors. Humans often increase storage capacity by constructing reservoirs and decrease it by draining wetlands. Humans often increase runoff quantities and velocities by paving areas and channelizing streamflow.

The total quantity of water available at any given time is an important consideration. Some human water users have an intermittent need for water. For

FIGURE 7.1   Life in Hur Al-Azim Wetland, Hoveyzeh, Iran (1960).

FIGURE 7.2   Karun River, Ahvaz, Iran (By Mohammad Albaji).

example, many farms require large quantities of water in the spring and no water at all in the winter. To supply such a farm with water, a surface water system may require a large storage capacity to collect water throughout the year and release it in a short period of time. Other users have a continuous need for water, such as a power plant that requires water for cooling. To supply such a power plant with water, a surface water system only needs enough storage capacity to fill in when average stream flow is below the power plant's need (Figure 7.3).

Nevertheless, over the long term the average rate of precipitation within a watershed is the upper bound for average consumption of natural surface water from that watershed.

Natural surface water can be augmented by importing surface water from another watershed through a canal or pipeline. It can also be artificially augmented from any of the other sources listed here; however, in practice the quantities are negligible. Humans can also cause surface water to be "lost" (i.e. become unusable) through pollution.

Brazil is the country estimated to have the largest supply of freshwater in the world, followed by Russia and Canada.

Under river flow

Throughout the course of the river, the total volume of water transported downstream will often be a combination of the visible free water flow together with a substantial

FIGURE 7.3 Lindeman Lake, Vancouver, Canada; (http://www.vancouvertrails.com/trails/lindeman-lake).

contribution flowing through sub-surface rocks and gravels that underlie the river and its floodplain called the hyporheic zone. For many rivers in large valleys, this unseen component of flow may greatly exceed the visible flow. The hyporheic zone often forms a dynamic interface between surface water and true ground-water receiving water from the groundwater when aquifers are fully charged and contributing water to groundwater when groundwaters are depleted. This is especially significant in karst areas where pot-holes and underground rivers are common.

Groundwater

Sub-surface water, or groundwater, is freshwater located in the pore space of soil and rocks. It is also water that is flowing within aquifers below the water table. Sometimes it is useful to make a distinction between sub-surface water that is closely associated with surface water and deep sub-surface water in an aquifer (sometimes called "fossil water").

Sub-surface water can be thought of in the same terms as surface water: inputs, outputs, and storage. The critical difference is that due to its slow rate of turnover, sub-surface water storage is generally much larger compared to inputs than it is for surface water. This difference makes it easy for humans to use sub-surface water unsustainably for a long time without severe consequences. Nevertheless, over the long term the average rate of seepage above a sub-surface water source is the upper bound for average consumption of water from that source.

The natural input to sub-surface water is seepage from surface water. The natural outputs from sub-surface water are springs and seepage to the oceans (Figure 7.4).

If the surface water source is also subject to substantial evaporation, a sub-surface water source may become saline. This situation can occur naturally under endorheic bodies of water, or artificially under irrigated farmland. In coastal areas, human use of a sub-surface water source may cause the direction of seepage to ocean to reverse which can also cause soil salinization. Humans can also cause sub-surface water to be "lost" (i.e. become unusable) through pollution. Humans can increase the input to a sub-surface water source by building reservoirs or detention ponds.

DESALINATION

Desalination is an artificial process by which saline water (generally seawater) is converted to freshwater. The most common desalination processes are distillation and reverse osmosis. Desalination is currently expensive compared to most alternative sources of water, and only a very small fraction of total human use is satisfied by desalination. It is only economically practical for high-valued uses (such as household and industrial uses) in arid areas.

Frozen water

Several schemes have been proposed to make use of icebergs as a water source, however to date this has only been done for novelty purposes. Glacier runoff is considered to be surface water.

The Himalayas, which are often called "The Roof of the World," contain some of the most extensive and rough high altitude areas on Earth as well as the greatest

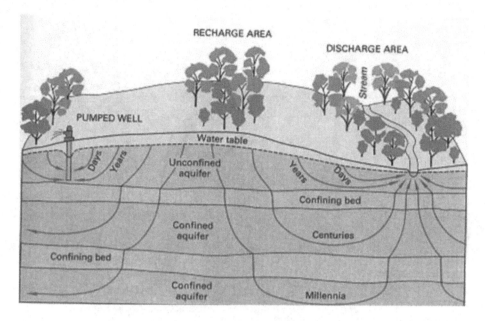

FIGURE 7.4 Sub-surface water travel time; (http://wi.water.usgs.gov/glpf/images/flow_ systems.jpg).

area of glaciers and permafrost outside of the poles. Ten of Asia's largest rivers flow from there and more than a billion people's livelihoods depend on them. To complicate matters, temperatures are rising more rapidly here than the global average. In Nepal, the temperature has risen by 0.6 degrees over the last decade, whereas the global warming has been around 0.7 over the last hundred years (Figure 7.5).

Uses of freshwater

Uses of freshwater can be categorized as consumptive and non-consumptive (sometimes called "renewable"). The use of water is consumptive if that water is not immediately available for another use. Losses to sub-surface seepage and evaporation are considered consumptive, as is water incorporated into a product (such as farm produce). Water that can be treated and returned as surface water, such as sewage, is generally considered non-consumptive if that water can be put to additional use.

Agricultural

It is estimated that 69% of worldwide water use is for irrigation, with 15–35% of irrigation withdrawals being unsustainable (Figures 7.6, 7.7, and Figure 7.8).

In some areas of the world irrigation is necessary to grow any crop at all, in other areas it permits more profitable crops to be grown or enhances crop yield. Various irrigation methods involve different trade-offs between crop yield, water

FIGURE 7.5   Adelie penguins on iceberg; (http://www.photosension.com/penguin-life/two-adelie-penguins-on-iceberg).

FIGURE 7.6   A wheat field, Shahid Chamran University of Ahvaz, Iran (By Mohammad Albaji).

FIGURE 7.7 A Lentils field, Shahid Chamran University of Ahvaz, Iran (By Mohammad Albaji).

FIGURE 7.8 Dew Irrigation, Shahid Chamran University of Ahvaz, Iran (By Mohammad Albaji).

consumption, and capital cost of equipment and structures. Irrigation methods such as furrow and overhead sprinkler irrigation are usually less expensive but are also typically less efficient, because much of the water evaporates, runs off, or drains below the root zone. Other irrigation methods considered to be more efficient include drip or trickle irrigation, surge irrigation, and some types of sprinkler systems where the sprinklers are operated near ground level. These types of systems, while more expensive, usually offer greater potential to minimize runoff, drainage, and evaporation. Any system that is improperly managed can be wasteful; all methods have the potential for high efficiencies under suitable conditions, appropriate irrigation timing, and management. One issue that is often insufficiently considered is salinization of sub-surface water.

Aquaculture is a small but growing agricultural use of water. Freshwater commercial fisheries may also be considered as agricultural uses of water, but have generally been assigned a lower priority than irrigation (see Aral Sea and Pyramid Lake).

As global populations grow, and as demand for food increases in a world with a fixed water supply, there are efforts underway to learn how to produce more food with less water, through improvements in irrigation methods and technologies, agricultural water management, crop types, and water monitoring.

Industrial

It is estimated that 22% of worldwide water use is industrial. Major industrial users include power plants, which use water for cooling or as a power source (i.e. hydroelectric plants), ore and oil refineries, which use water in chemical processes, and manufacturing plants, which use water as a solvent.

The portion of industrial water usage that is consumptive varies widely, but as a whole is lower than agricultural use (Figure 7.9).

Water is used in power generation. Hydroelectricity is electricity obtained from hydropower. Hydroelectric power comes from water driving a water turbine connected to a generator. Hydroelectricity is a low-cost, non-polluting, renewable energy source. The energy is supplied by the sun. The heat from the sun evaporates water, which condenses as rain in higher altitudes, from where it flows down.

Pressurized water is used in water blasting and water jet cutters. Also, very high-pressure water guns are used for precise cutting. It works very well, is relatively safe, and is not harmful to the environment. It is also used in the cooling of machinery to prevent overheating, or prevent saw blades from overheating.

Water is also used in many industrial processes and machines such as the steam turbine and heat exchanger, in addition to its use as a chemical solvent. Discharge of untreated water from industrial uses is pollution. Pollution includes discharged solutes (chemical pollution) and discharged coolant water (thermal pollution). Industry requires pure water for many applications and utilizes a variety of purification techniques both in water supply and discharge.

FIGURE 7.9 Karun-3 power plant, Khuzestan, Iran: (http://www.industcards.com/hydro-iran.htm).

HOUSEHOLD

It is estimated that 8% of worldwide water use is for household purposes. These include drinking water, bathing, cooking, sanitation, and gardening. Basic household water requirements have been estimated by Peter Gleick at around 50 liters per person per day, excluding water for gardens. Drinking water is water that is of sufficiently high quality so that it can be consumed or used without risk of immediate or long-term harm. Such water is commonly called potable water. In most developed countries, the water supplied to households, commerce and industry is all of drinking water standard even though only a very small proportion is actually consumed or used in food preparation.

RECREATION

Recreational water use is usually a very small but growing percentage of total water use. Recreational water use is mostly tied to reservoirs. If a reservoir is kept fuller than it would otherwise be for recreation, then the water retained could be categorized as recreational usage. Release of water from a few reservoirs is also timed to enhance whitewater boating, which also could be considered a recreational usage. Other examples are anglers, water skiers, nature enthusiasts, and swimmers (Figures 7.10, Figure 7.11, and Figure 7.12).

FIGURE 7.10 Beautiful water nature ,The Greek shipwrecked, Kish Island, Iran (By Mohammad Albaji).

Recreational usage is usually non-consumptive. Golf courses are often targeted as using excessive amounts of water, especially in drier regions. It is, however, unclear whether recreational irrigation (which would include private gardens) has a noticeable effect on water resources. This is largely due to the unavailability of reliable data. Additionally, many golf courses utilize either primarily or exclusively treated effluent water, which has little impact on potable water availability (Figure 7.13).

Some governments, including the Californian Government, have labeled golf course usage as agricultural in order to deflect environmentalists' charges of wasting water. However, using the above figures as a basis, the actual statistical effect of this reassignment is close to zero. In Arizona, an organized lobby has been established in the form of the Golf Industry Association, a group focused on educating the public on how golf impacts the environment.

Recreational usage may reduce the availability of water for other users at specific times and places. For example, water retained in a reservoir to allow boating in the late summer is not available to farmers during the spring planting season. Water released for whitewater rafting may not be available for hydroelectric generation during the time of peak electrical demand.

FIGURE 7.11 Boating in Turkey (By Mohammad Albaji).

ENVIRONMENTAL

Explicit environmental water use is also a very small but growing percentage of total water use. Environmental water usage includes artificial wetlands, artificial lakes intended to create wildlife habitat, fish ladders, and water releases from reservoirs timed to help fish spawn.

Like recreational usage, environmental usage is non-consumptive but may reduce the availability of water for other users at specific times and places. For example, water released from a reservoir to help fish spawn may not be available to farms upstream.

WATER STRESS

The concept of water stress is relatively simple: According to the World Business Council for Sustainable Development, it applies to situations where there is not enough water for all uses, whether agricultural, industrial, or domestic. Defining

FIGURE 7.12   Swimming, Muscat, Oman (By Mohammad Albaji).

thresholds for stress in terms of available water per capita is more complex, however, entailing assumptions about water use and its efficiency. Nevertheless, it has been proposed that when annual per capita renewable freshwater availability is less than 1,700 cubic meters, countries begin to experience periodic or regular water stress. Below 1,000 cubic meters, water scarcity begins to hamper economic development and human health and well-being (Figure 7.14).

POPULATION GROWTH

In 2000, the world population was 6.2 billion. The UN estimates that by 2050 there will be an additional 3.5 billion people with most of the growth in developing countries that already suffer water stress. Thus, water demand will increase unless there are corresponding increases in water conservation and recycling of this vital resource.

EXPANSION OF BUSINESS ACTIVITY

Business activity ranging from industrialization to services such as tourism and entertainment continues to expand rapidly. This expansion requires increased water services including both supply and sanitation, which can lead to more pressure on water resources and natural ecosystems.

FIGURE 7.13   Drinking water (By Mohammad Albaji).

Rapid Urbanization

The trend toward urbanization is accelerating. Small private wells and septic tanks that work well in low-density communities are not feasible within high-density urban areas. Urbanization requires significant investment in water infrastructure in order to deliver water to individuals and to process the concentrations of wastewater – both from individuals and from business. These polluted and contaminated waters must be treated or they pose unacceptable public health risks.

In 60% of European cities with more than 100,000 people, groundwater is being used at a faster rate than it can be replenished. Even if some water remains available, it costs more and more to capture it.

Climate Change

Climate change could have significant impacts on water resources around the world because of the close connections between the climate and hydrological cycle. Rising

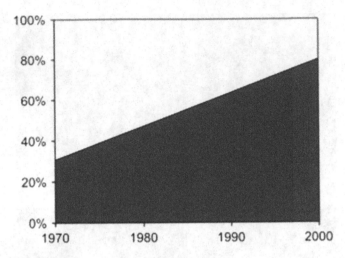

FIGURE 7.14 Best estimate of the share of people in developing countries with access to drinking water 1970–2000; (https://en.wikipedia.org/wiki/File:Access_to_drinking_water_in_third_world.svg).

temperatures will increase evaporation and lead to increases in precipitation, though there will be regional variations in rainfall. Overall, the global supply of freshwater will increase. Both droughts and floods may become more frequent in different regions at different times, and dramatic changes in snowfall and snowmelt are expected in mountainous areas. Higher temperatures will also affect water quality in ways that are not well understood. Possible impacts include increased eutrophication. Climate change could also mean an increase in demand for farm irrigation, garden sprinklers, and perhaps even swimming pools.

DEPLETION OF AQUIFERS

Due to the expanding human population, competition for water is growing such that many of the world's major aquifers are becoming depleted. This is due both to direct human consumption as well as agricultural irrigation by groundwater. Millions of pumps of all sizes are currently extracting groundwater throughout the world. Irrigation in dry areas such as northern China and India is supplied by groundwater, and is being extracted at an unsustainable rate. Cities that have experienced aquifer drops between 10 and 50 meters include Mexico City, Bangkok, Manila, Beijing, Madras, and Shanghai.

POLLUTION AND WATER PROTECTION

Water pollution is one of the main concerns of the world today. The governments of numerous countries have strived to find solutions to reduce this problem. Many pollutants threaten water supplies, but the most widespread, especially in developing countries, is the discharge of raw sewage into natural waters; this method of sewage disposal is the most common method in underdeveloped countries but also is prevalent in quasi-developed countries such as China, India, and Iran. Sewage, sludge, garbage, and even toxic pollutants are all dumped into the water. Even if sewage is treated, problems still arise. Treated sewage forms sludge, which may be placed in landfills, spread out on land, incinerated, or dumped at sea. In addition to sewage, nonpoint source pollution such as agricultural runoff is a significant source of pollution in some parts of the world, along with urban stormwater runoff and chemical wastes dumped by industries and governments (Figure 7.15).

WATER AND CONFLICT

The only known example of an actual inter-state conflict over water took place between 2500 and 2350 BC between the Sumerian states of Lagash and Umma. Yet, despite the lack of evidence of international wars being fought over water alone, water has been the source of various conflicts throughout history. When water scarcity causes political tensions to arise, this is referred to as water stress. Water stress has led most often to conflicts at local and regional levels. Using a purely quantitative methodology, Thomas Homer-Dixon successfully correlated water scarcity and scarcity of available arable lands to an increased chance of violent conflict.

FIGURE 7.15 Discharge of raw sewage to Karun River, Ahvaz, Iran (By Mohammad Albaji).

Water stress can also exacerbate conflicts and political tensions which are not directly caused by water. Gradual reductions over time in the quality and/or quantity of freshwater can add to the instability of a region by depleting the health of a population, obstructing economic development, and exacerbating larger conflicts.

Conflicts and tensions over water are most likely to arise within national borders, in the downstream areas of distressed river basins. Areas such as the lower regions of China's Yellow River or the Chao Phraya River in Thailand, for example, have already been experiencing water stress for several years. Additionally, certain arid countries which rely heavily on water for irrigation, such as China, India, Iran, and Pakistan, are particularly at risk of water-related conflicts. Political tensions, civil protest, and violence may also occur in reaction to water privatization. The Bolivian Water Wars of 2000 are a case in point.

WORLD WATER SUPPLY AND DISTRIBUTION

Food and water are two basic human needs. However, global coverage figures from 2002 indicate that of every 10 people:

- roughly 5 have a connection to a piped water supply at home (in their dwelling, plot, or yard);
- 3 make use of some other sort of improved water supply such as a protected well or public standpipe;
- 2 are unserved;
- In addition, 4 out of every 10 people live without improved sanitation.

At Earth Summit 2002 governments approved a Plan of Action to:

- Halve by 2015 the proportion of people unable to reach or afford safe drinking water. The Global Water Supply and Sanitation Assessment 2000 Report (GWSSAR) defines "Reasonable access" to water as at least 20 liters per person per day from a source within one kilometer of the user's home.
- Halve the proportion of people without access to basic sanitation. The GWSSR defines "Basic sanitation" as private or shared but not public disposal systems that separate waste from human contact.

As the picture shows, in 2025, water shortages will be more prevalent among poorer countries where resources are limited and population growth is rapid, such as the Middle East, Africa, and parts of Asia. By 2025, large urban and peri-urban areas will require new infrastructure to provide safe water and adequate sanitation. This suggests growing conflicts with agricultural water users, who currently consume the majority of the water used by humans.

Generally speaking the more developed countries of North America, Europe, and Russia will not see a serious threat to water supply by the year 2025; not only because of their relative wealth, but more importantly their populations will be better aligned with available water resources. North Africa, the Middle East, South

Africa, and northern China will face very severe water shortages due to physical scarcity and a condition of overpopulation relative to their carrying capacity with respect to water supply. Most of South America, Sub-Saharan Africa, Southern China, and India will face water supply shortages by 2025; for these latter regions, the causes of scarcity will be economic constraints to developing safe drinking water, as well as excessive population growth.

One point six billion people have gained access to a safe water source since 1990. The proportion of people in developing countries with access to safe water is calculated to have improved from 30% in 1970 to 71% in 1990, 79% in 2000 and 84% in 2004. This trend is projected to continue.

ECONOMIC CONSIDERATIONS

Water supply and sanitation require a huge amount of capital investment in infrastructure such as pipe networks, pumping stations and water treatment works. It is estimated that Organization for Economic Co-operation and Development (OECD) nations need to invest at least USD 200 billion per year to replace aging water infrastructure to guarantee supply, reduce leakage rates and protect water quality.

International attention has focused upon the needs of the developing countries. To meet the Millennium Development Goals targets of halving the proportion of the population lacking access to safe drinking water and basic sanitation by 2015, the current annual investment on the order of USD 10 to USD 15 billion would need to be roughly doubled. This does not include investments required for the maintenance of existing infrastructure.

Once infrastructure is in place, operating water supply and sanitation systems entails significant ongoing costs to cover personnel, energy, chemicals, maintenance, and other expenses. The sources of money to meet these capital and operational costs are essentially either user fees, public funds, or some combination of the two.

But this is where the economics of water management start to become extremely complex as they intersect with social and broader economic policy. Such policy questions are beyond the scope of this article, which has concentrated on basic information about water availability and water use. They are, nevertheless, highly relevant to understanding how critical water issues will affect business and industry in terms of both risks and opportunities.

BUSINESS RESPONSE

The World Business Council for Sustainable Development in its H2O Scenarios engaged in a scenario building process to:

- Clarify and enhance understanding by business of the key issues and drivers of change related to water.
- Promote mutual understanding between the business community and non-business stakeholders on water management issues.
- Support effective business action as part of the solution to sustainable water management.

It concludes that:

- Business cannot survive in a society that thirsts.
- One does not have to be in the water business to have a water crisis.
- Business is part of the solution, and its potential is driven by its engagement.
- Growing water issues and complexity will drive up costs.

EXERCISES

A. **Read each statement and decide whether it is true or false. Write "T" for true and "F" for false statements.**

TF1. Water resources have been used for agricultural, industrial, household, recreational, and environmental activities.

TF2. Surface water is naturally lost through discharge to river, lake, or freshwater wetland.

TF3. Desalination is a natural process by which saline water is converted to freshwater.

TF4. The only natural input to any surface water system is precipitation.

TF5. A use of water is consumptive if that water is immediately available for another use.

B. **Circle a, b, c, or d which best completes the following items.**

1. 97% of water on the earth is ...., and only 3% is .... of which slightly over two-thirds is frozen in glaciers and polar ice caps.
   a. saltwater - useable water
   b. freshwater - saltwater
   c. saltwater - freshwater
   d. freshwater - polluted water

2. The critical difference is that due to its slow rate of turnover, ... storage is generally much larger compared to inputs than it is for ....
   a. sub-surface water - surface water.
   b. surface water - sub-surface water
   c. sub-surface water - fossil water
   d. surface water- fossil water

3. It is estimated that 70% of worldwide water use is for ....
   a. recreational.
   b. household.
   c. industrial.
   d. agricultural.

4. .... applies to situations where there is not enough water for all uses, whether agricultural, industrial, or domestic.
   a. Water useless
   b. Water scarcity
   c. Water stress
   d. Waste water

5. .... could have significant impacts on water resources around the world because of the close connections between the climate and ....
   a. Hydrological cycle - climate change
   b. Climate change - hydrological cycle
   c. Climate cycle - hydrological change
   d. Hydrological change - climate cycle

C. **Match the sentence halves in Column I with their appropriate halves in Column II. Insert the letters a, b, c ...in the parentheses provided. There are more sentence halves in Column II than required.**

Column I	Column II
1. Water demand already	() **a.** have an intermittent need for water.
2. Humans often increase storage capacity	() **b.** have been estimated at around 50 liters per person per day.
3. Some human water users	() **c.** water that is flowing within aquifers below the water table.
4. Sub-surface water can be thought of	() **d.** have not a large and sometimes devastating impact on surface water.
5. Basic household water requirements	() **e.** by constructing reservoirs and decreasing them by draining wetlands.
	() **f.** is water that is of sufficiently high quality so that it can be consumed or used without risk of immediate or long-term harm.
	() **g.** exceeds supply in many parts of the world.
	() **h.** in the same terms as surface water: inputs, outputs, and storage.

D. **Cross out the word or words that make each statement false, and write the word or words that make each statement true in the blank.**
   1. Freshwater is a renewable resource, yet the world's supply of clean, freshwater is steadily increasing.
      ....
   2. The natural output to sub-surface water is seepage from surface water. The natural inputs from sub-surface water are springs and seepage to the oceans.
      ....

3. Losses to sub-surface seepage and evaporation are considered none consumptive, as is water incorporated into a product.
   ....

4. In addition to sewage, point source pollution such as agricultural runoff is a significant source of pollution in some parts of the world.
   ....

E. **Give answers to the following questions.**
   1. What are the water resources?
   2. How much water resources are used in agricultural, industrial, household, recreational, and environmental activities, respectively?
   3. What are the major industrial water users?

F. **For each word on the left, there are three meanings provided. Put a check mark (√) next to the choice which has the closest meaning to the word given.**

1. Desalination	convert to saline water	convert to freshwater	convert to clean water
2. Fossil water	deep sub-surface water	sub-surface water	groundwater
3. Renewable	unconsumptive	consumptive	non-consumptive
4. Fish ladders	fish path	fish way	fish spillway
5. Household water	drinking water	agricultural water	recreational water

G. **Fill in the blanks with the appropriate words from the following list.**
   forests   glaciers   green water   groundwater discharge
   atmosphere
   freshwater   evapotranspiration   water cycle

The world's water exists naturally in different forms and locations: in the air, on the surface, below the ground, and in the oceans. Just 2.5% of the Earth's water is ...., and most is frozen in .... and ice sheets. About 96% of all liquid freshwater can be found underground. The remaining small fraction is on the surface or in the air. Knowing how water cycles through the environment can help in determining how much water is available in different parts of the world. The Earth's .... is the global mechanism by which water moves from the air to the Earth (precipitation) and eventually back to the .... (Evaporation).

The principal natural components of this cycle are precipitation, infiltration into the soil, runoff on the surface, ... to surface waters and the oceans, and .... from water bodies, the soil, and plants. "Blue water "— the water in rivers, lakes, and aquifers— can be distinguished from "............... " — which feeds plants and crops, and which is subsequently released into the air. This distinction may help managers focus on those areas which green water feeds and passes through, such as farms, ..., and wetlands.

H. **Read this passage and then circle a, b, c, or d which best completes the following items.**

Around the world, human activity and natural forces are reducing available water resources. Although public awareness of the need to better manage and protect water has grown over the last decade, economic criteria and political considerations still tend to drive water policy at all levels. Science and best practice are rarely given adequate consideration. Pressures on water resources are increasing mainly as a result of human activity – namely urbanization, population growth, increased living standards, growing competition for water, and pollution. These are aggravated by climate change and variations in natural conditions.

The main purpose of this passage is ....

a. reducing available water resources.

b. pressures on water resources.

c. the effect of climate change on water resources.

d. protection of water resources.

I. **Translate the following passage into your mother language. Write your translation in the space provided.**

About 10% of the Earth's freshwater that is neither frozen nor underground is found in the atmosphere. Precipitation, in the form of rain or snow, for instance, is an important form of available freshwater. About 40% of precipitation has previously evaporated from the oceans; the rest from land. The amount of precipitation varies greatly around the world, from less than 100 mm a year in desert climates to over 3400 mm a year in tropical settings.

In temperate climates, about a third of precipitation returns to the atmosphere through evaporation, a third filters into the ground and replenishes groundwater and the remainder flows into water bodies. The drier the climate, the higher the proportion of precipitation that returns to the atmosphere and the lower the proportion that replenishes groundwater.

A large part of the freshwater that returns to the atmosphere passes through soil and plants. Reliable figures are available only for some regions. Soil moisture is important for plant growth. Finding out how much moisture soil contains is important for such activities as farming and "river-flow forecasting ", and for understanding climate and natural and water systems. Satellite data are increasingly complementing measurements of soil moisture taken

on the ground to provide a broader and more up-to-date picture to
decision-makers.

_____
_____
_____
_____
_____
_____
_____
_____
_____
_____
_____
_____
_____
_____
_____

J. **Copy the technical terms and expressions used in this lesson.
Then find your mother language equivalents of those terms and
expressions and write them in the spaces provided.**

Technical term   **Mother language equivalent**

....
....
....
....
....
....
....
....
....
....

BIBLIOGRAPHY

Addressing Our Global Water Future PDF from the Center for Strategic and International
    Studies (CSIS) / Sandia National Laboratories.
American Water Resources Association.
Ancient Irrigation from the University of California, Geology Department.
Canadian Water Resources Association.

FAO Water Portal Food and Agriculture Organization of the United Nations.
http://beldibi.biz/en/rafting-turkey
http://edition.presstv.ir/detail.fa/167369.html
https://en.wikipedia.org/wiki/File:Access_to_drinking_water_in_third_world.svg
https://en.wikipedia.org/wiki/ Water resources
http://picphotoz.com/farm-with-us
http://wi.water.usgs.gov/glpf/images/flow_systems.jpg
http://www.chemistryexplained.com/Va-Z/Water-Pollution.html
http://www.h₂o₂.com/municipal-applications/drinking-water-
        treatment
http://www.industcards.com/hydro-iran.htm
http://www.photosension.com/penguin-life/two-adelie-penguins-on-iceberg
http://www.techstart.org/docs/ogpc_payment.pdf
http://www.vancouvertrails.com/trails/lindeman-lake
Institute for Water Resources - USACE.
International Water Resources Association.
Mining Water from the University of California, Geology Department.
Nikravesh, Ardakanian and Alemohammad, Institutional Capacity Development of Water
        Resources Management in Iran.
Selected World Water Data.
The United Nations World Water Development Report 2 Section 2: Changing Natural
        Systems, Chapter 4, Part 2. Nature, Variability and Availability, p.124.
UN World Water Development Report.
Water and Cities: Acting on the Vision.
Water and the Future of Life on Earth.
Water Resources of the United States.
Water Resource Research Center.
World Water Supply and Demand: 1995 to 2025 from the International Water Management
        Institute.

GLOSSARY OF TERMS

Aquifer: A geologic formation that will yield water to a well in sufficient quantities
        to make the production of water from this formation feasible for beneficial
        use; permeable layers of underground rock or sand that hold or transmit
        groundwater below the water table.

Desalination: The process of salt removal from sea or brackish water.

Evaporation: The change by which any substance is converted from a liquid state
        and carried off in vapor. Compare condensation, sublimation.

Freshwater: Water containing less than 1,000 parts per million (ppm) of dissolved
        solids of any type.

Glacier: A huge mass of land ice that consists of recrystallized snow and moves
        slowly downslope or outward.

Groundwater: Water within the earth that supplies wells and springs; water in the
        zone of saturation where all openings in rocks and soil are filled, the upper
        surface of which forms the water table.

Reservoir: A pond, lake, tank basin, or other space natural or created, which is used
        for the storage regulation, and control of water for recreation, power, flood
        control, or consumption.

Runoff: That portion of precipitation that flows off the surface of a drainage area after accounting for all abstractions such as interception, evaporation, infiltration, and surface storage.

Seepage: Percolation of water through the soil from unlined canals, ditches, laterals, watercourses, or water storage facilities.

Sewage: Household and commercial wastewater that contains human waste.

Surface water: Water that flows in streams and rivers and in natural lakes, in wetlands, and in reservoirs constructed by humans.

Water pollution: Degradation of a body of water by a substance or condition to such a degree that the water fails to meet specified standards or cannot be used for a specific purpose.

Watershed: Land area from which water drains toward a common watercourse in a natural basin.

8 Aquatic ecosystem

READING FOR COMPREHENSION

Aquatic ecosystem is an ecosystem in a water body environment in which communities of organisms including plants and animals, live, interact, and are dependent on each other and on their environment. Aquatic ecosystems connect people, land, and wildlife through the water. Aquatic ecosystems are generally classified into two categories based on the number of dissolved compounds, especially salt in water: the marine ecosystem and the freshwater ecosystem (Figure 8.1).

MARINE ECOSYSTEM

Marine ecosystems are the most extensive ecosystems that cover over 70% of the earth's surface and produce 32% of the world's net primary production. Approximately 97% of the planet's water is marine water with a high range of salinity in different marine ecosystems. For example, seawater has an average salinity of 35 parts per thousand of water (ppt) and nearly 85% of the dissolved materials in seawater are sodium and chlorine. Marine organisms are a certain feature of marine ecosystems. Biological community of organisms in Marine ecosystem should have conformity to both conditions of continuous changing or fixed amount of salt and cannot move successfully from one to the other. Animals in marine ecosystems range from microscopic zooplankton through fish of all sizes to marine mammals, including seals, whales, and manatees.

MARINE ECOSYSTEM CLASSIFICATION

Coastal

A coastal ecosystem is an area where land and water come together. Natural processes such as weather and sea level change cause the erosion, accretion, and resculpturing of coasts, which makes the coastal ecosystem a dynamic environment with constant changes. The main factors in deposition and erosion along coastlines are waves, tides, and currents. In addition, the hardness of the rocks that make the coasts causes difference in the coastlines shape. The hardness of the rocks has an inverse relation to the probability of erosion. Coastal ecosystems in division in different types including intertidal, sandy shores, rocky shores, mudflats, mangrove and salt marshes, estuaries, kelp forests, sea grass meadows, coral reefs (Figure 8.2).

Oceanic Zone

Oceanic zones, also known as open oceans or pelagic ecosystems, are the areas away from the coastal boundaries and above the bed of the sea. These enormous ecosystems cover large parts of earth's surface by five major oceans including

DOI: 10.1201/9781003293507-8

FIGURE 8.1 Aquatic ecosystem; (https://theblogaquatic.org/aquatic-ecosystem-facts/).

FIGURE 8.2 Coastal ecosystem, Muscat, Oman (By Mohammad Albaji).

FIGURE 8.3   Southern Ocean; (https://www.scientificamerican.com/article/oceans-are-warming-faster-than-predicted/).

Pacific Ocean, Indian Ocean, Arctic Ocean, Atlantic Ocean, and Southern (Antarctic) Ocean, each one of these oceans have their own unique species and features, although they are connected (Figure 8.3).

The pelagic ecosystems classifying into five different zones based on the water depth:

Epipelagic zone (Photic zone) ranges from the sea surface to depth about 200 meters. In this zone, the enough light is available for photosynthesis and approximately contains all primary production of the ocean, which causes concentration of plants and animals. For instance, sharks, dolphins, tuna, floating seaweed, jellyfish, and plankton are living organisms of this zone. Organisms of the epipelagic zone may contact with the sea surface.

Mesopelagic zone is the zone between the bottoms of the epipelagic in the 200 meters depth to about 1,000 meters and is much larger than the epipelagic. Most vertebrates of earth live in mesopelagic zone. Many species of fishes and invertebrates of mesopelagic zone move up to shallower depths in epipelagic zone for feeding, but only at night.

Bathypelagic zone stretches from 1,000 depth to 4,000 meters. The bathypelagic is the largest ecosystem on earth that is much larger than the mesopelagic and 15 times the size of the epipelagic. A complete lack of sunlight determines the upper bound of this zone and organisms live in permanent complete darkness, however, some light of bioluminescence can interrupt the darkness. Animals of this zone, feeding by the marine snow that falling from the upper zones, or by hunting the other inhabitants of this zone.

Abyssopelagic zone extends from a water depth of 4,000 meters to 6,000 meters that known as a relative lack of life because of cold temperatures, high pressures, and complete darkness. The number of living creatures in this zone is very small and many of them are transparent and eyeless due to the total lack of light. Several species of squid that are found in this zone are including echinoderms (such as the basket star, swimming cucumber, and the sea pig) and marine arthropods (such as the sea spider).

Hadopelagic zone is the deepest zone in the ocean at more than 6,000 m depth. This zone is found in deep, wide trenches filled by the open water. The maximum known depth on the earth, known as the Challenger Deep, is about 10,984 meters and located in the western Pacific Ocean (Figure 8.4).

Marine Pollution

Marine pollution is a serious concern that has a destructive effect on the environment, organisms healthy, and world economy. There are many methods to classify the marine pollution based on the path of entry and the type of pollution. Generally, pollution is categorized into two classes depending on the pathway of entry: point source or nonpoint source pollution. Point source pollution is a single, identifiable, localized source of the pollution that occurs specially in developing nations. Directly discharging sewage and industrial waste into the ocean is the most important point source of pollution. These sources are called point sources because in mathematical modeling they can be approximated as a mathematical point to simplify analysis. Nonpoint source pollution is resulting from diffuse sources such as agricultural runoff and wind-blown debris. This pollution can be difficult to regulate.

Another method to classify the pollution is considering their type. There are two main types of pollution: chemicals and trash. Chemical or nutrient pollution occurs when human actions, particularly using fertilizer in agriculture, produce the chemicals runoff that flows into the ocean through waterways. High concentration of chemicals such as nitrogen and phosphorus, increases the growth of algal blooms in the coastal ocean, which can be toxic to wildlife and humans, hurt tourism industries, and local fishing. Marine trash contains all manufactured products that 80% of which come from lands sources. Some factors such as Littering, storm winds, and poor waste management promote the accumulation of debris. Most common marine debris are different types of plastic such as shopping bags, beverage bottles, bottle caps, food wrappers, fishing gear, and cigarette filters. Many studies have been conducted to estimate the amount of plastic in the oceans. For instance, a scientific study estimated the mass of plastic in the oceans was nearly 150 million tons in 2016 and predicted that this amount would increase to 250 million tons by 2025. The main problem with plastics is their biodegradable way that is not like many other substances. Plastics will photodegrade on exposure to the sun but only under dry conditions, and water prevents this process (Figure 8.5).

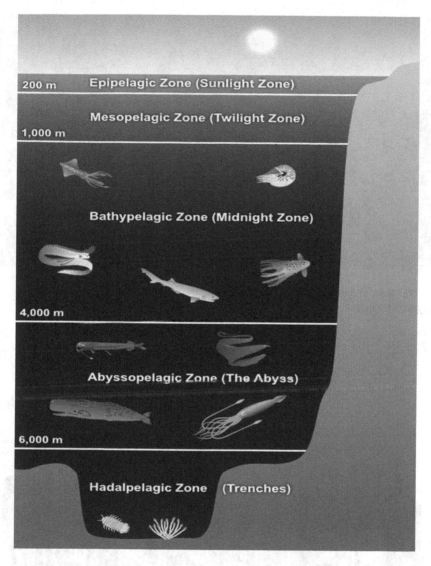

FIGURE 8.4 Pelagic zone divisions; (http://www.seasky.org/deep-sea/ocean-layers.html).

Despite the long history of marine pollution, there was no important international law for controlling them until the 20th century and most scientists believed that the oceans, due to their large size, had an unlimited ability to dilute and reduce pollution. Nevertheless, in the last decades, the international community has accepted the reduction of ocean pollution as a priority, therefore United Nations considered laws, policies, and treaties such as International Convention for the Prevention of Pollution from Ships, Law of the Sea Convention, Sustainable Development Goal 14. In addition, according to a report from the United Nations in 2018, more than 60 countries have enacted regulations to limit or ban the use of disposable plastic items (Figure 8.6).

FIGURE 8.5 Accumulation of marine pollution; (https://www.thegef.org/news/ocean-everyones-business).

FIGURE 8.6 Our Ocean Conference in Indonesia, 2018; (https://ourocean2018.org/).

FRESHWATER ECOSYSTEM

Freshwater ecosystems are a subset of aquatic ecosystems that contain 0.009% of total planet water and cover 0.78% of the earth's surface with rivers, streams, lakes, pools, ponds, wetlands, and even groundwater. Freshwater ecosystems are habitats for a high variety of animals including insects, amphibians, and fish. Some studies estimate the number of different species of fish living in freshwater to be at least 45,000, accounting for more than 40% of the world's known fish species.

FRESHWATER ECOSYSTEM CLASSIFICATION

Lentic

Lentic ecosystems are standing or slow-moving water habitats including pools, ponds, and lakes. Actually, the lentic word means stationary or relatively still water. The greatest energy source in lentic ecosystem is light obtained from the solar energy that is necessary for the photosynthesis process. Due to the high population of poikilothermic animals in biota of the lentic ecosystems, temperature is another main abiotic factor in this ecosystem. Temperature regimes can vary greatly in these systems, either on a daily or seasonal basis, depending on the region in which they are located. Lake ecosystems, based on one common system, are divided into three zones including the littoral zone, the photic zone (open water zone), and the aphotic zone (deep water zone). Lentic ecosystems are habitats for algae, crabs, shrimps, amphibians (frogs and salamanders, etc.), reptiles (alligators, water snakes, etc.), and plants (rooted and floating-leaved) (Figure 8.7).

Lotic

They are the rapidly flowing waters that refer to the streams and rivers. Where the river bed has a steep slope or the current is turbulent with high velocity, water has

FIGURE 8.7 Dez Dam Lake in Iran; (https://www.eligasht.com/Blog/wp-content/uploads/2018/06/01-6.jpg).

more concentrations of dissolved oxygen usually which led to supporting greater biodiversity compared to the slow-moving water of pools. Running waters are unique among aquatic ecosystems for studying ecology because they have some advantages including unidirectional flow, continuous physical change, high degree of spatial and temporal heterogeneity at all scales, high variability between lotic systems, specialized biota to live with flow conditions. Lotic ecosystems are home to many species of insects (beetles, mayflies, and stoneflies), fishes (trout, eel, minnow, etc.), and mammals (beavers, river dolphins, and otters) (Figure 8.8).

Wetlands

Wetlands are defined as areas where the soil is saturated or inundated for at least part of the time. Primary factor that produces wetlands is flooding. The flooding duration or prolonged soil saturation by groundwater determines the type of the created wetland: swamp, marsh, fen, or bog. Wetlands are distinguished from other landforms or water bodies by vegetation of aquatic plants. The wetlands are advantageous for some purposes including water storage, water treatment, replacement of groundwater, flood control, stabilization of shoreline, and protection against storms (Figure 8.9).

FIGURE 8.8 Karun River, Ahvaz, Iran (By Mohammad Albaji).

FIGURE 8.9 Hur Al-Azim Wetland in Hoveyzeh, Iran (By Mohammad Albaji).

EXERCISES

A. **Read each statement and decide whether it is true or false. Write "T" for true and "F" for false statements.**

TF1. Aquatic ecosystems are located in both water bodies and drylands.

TF2. Most of the dissolved materials in seawater are sodium and chlorine.

TF3. The coastal ecosystems are stable environments without significant changes.

TF4. Mesopelagic zone stretches from the bottom of the epipelagic zone in the 200 meters depth to 4,000 meters.

TF5. Flooding is the main factor in producing wetlands.

B. **Circle a, b, c, or d which best completes the following items.**

1. .... must have the ability to adapt to both conditions of continuous changing or stable amount of salt.
   a. Fishes
   b. Marine organisms
   c. Freshwater organisms
   d. Humans in coastlines

2. Coastal ecosystems classify into several classes such as intertidal, sandy shores, rocky shores, mudflats, coral reefs and ....
   a. open oceans.
   b. rivers.
   c. wetlands.
   d. estuaries.

3. Directly discharging sewage and industrial waste into the ocean are identifiable, localized pollution that is called ....
   a. point source pollution.
   b. fertilizer pollution.
   c. nonpoint source pollution.
   d. debris.

4. Today, the international community has agreed that ....
   is a very important issue.
   a. take a more careful study on plastics pollution in oceans
   b. decreasing the pollution in the oceans
   c. the development of ocean-based industries
   d. make more use of the oceans

5. .... are slow-moving or still waters including pools, ponds, and lakes.
   a. Lotic ecosystems
   b. Freshwater ecosystems
   c. Lentic ecosystems
   d. Marine ecosystems

C. **Match the sentence halves in Column I with their appropriate halves in Column II. Insert the letters a, b, c ... in the parentheses provided. There are more sentence halves in Column II than required.**

Column I	Column II
1. Abiotic component	() **a.** the deepest known point in the Earth's seabed.
2. Saturation	() **b.** the alteration of materials by light.
3. Primary production	() **c.** non-living part of an ecosystem that shapes its environment.
4. Photodegradation	() **d.** when all of the voids, spaces, and cracks are filled with water.
5. Challenger Deep	() **e.** the synthesis of organic compounds.
	() **f.** decreasing the concentration of a solute in a solution.
	() **g.** where land and water come together.
	() **h.** deep water zone.

D. **Cross out the word or words that make each statement false, and write the word or words that make each statement true in the blank.**
   1. Aquatic ecosystems are divided into marine ecosystems and freshwater ecosystems based on water depth.

   ....
   2. The enormous ecosystems cover large parts of earth's surface by five major oceans called lotic ecosystems.

   ....
   3. Almost all of the primary production in the ocean accrue in bathypelagic zone because there is enough light for photosynthesis.

   ....
   4. Lotic ecosystems have low flow velocity and gentle fall in the bed river that result in more concentrations of dissolved oxygen usually.

   ....
E. **Give answers to the following questions.**
   1. What is the main difference between the marine ecosystem and the freshwater ecosystem?
   2. Why the coastal ecosystem is a dynamic environment?
   3. Why the number of living creatures in Abyssopelagic zone is very small and many of them are eyeless?
   4. How the United Nations helps to reduction of ocean pollution?
F. **For each word on the left, there are three meanings provided. Put a check mark (√) next to the choice which has the closest meaning to the word given.**

1. Oceanic zone	aphotic zone	pelagic zone	coastal zone
2. Biodegradation	erosion	decomposition	accumulation
3. Dilution	equation	infiltration	degradation
4. Mudflats	rocky shores	tidal flats	coral reefs
5. Seawater	pure water	freshwater	salt water

G. **Fill in the blanks with the appropriate words from the following list.**

abiotic component   lentic   marine pollution   marine water
coastal   oceans   lotic   Aquatic ecosystem

Ecosystems have two types including Terrestrial ecosystem and .... This ecosystem is a water-based environment, in which living organisms interact with .... of the environment.

Generally, there are two different classes of aquatic ecosystem namely marine ecosystem and freshwater ecosystem. Approximately 97% of all the water on earth are .... in .... seas and saline groundwater. Environmental threats to marine ecosystems are including .... unsustainable exploitation of marine resources, building on .... zones and climate change. Freshwaters contain about 2.5–2.75% of planet water including 1.75–2% frozen in glaciers, ice, and snow, 0.5–0.75% as fresh groundwater and soil moisture, and less than 0.01% of it as surface water in .... ecosystems such as lakes, ... ecosystems such as rivers, and wetlands.

H. **Read this passage and then circle a, b, c, or d which best completes the following items.**

Coral reefs are one of the most known underwater marine ecosystems and formed of coral skeletons from mostly intact coral colonies. Coral reefs furnish benefits for tourism and fisheries, and protect the coastline, which results in preventing loss of life and property damages. Reefs can decrease waves energy as same as or even better than designed structures for coastal defense such as breakwaters. In some cases, reduction of waves energy by coral reefs is about 97%. Many small islands would disappear without protective reefs. In addition, coastlines protected by coral reefs are more stable in erosion parameters compared to others. Various animals that live on coral reefs are including fishes (over 4,000 species), cnidarians, worms, seabirds, sponges, crustaceans, mollusks, sea snakes, sea squirts, and sea turtles.

The artificial structures designed ....

a. never can be useful for coastal protection compared to coral reefs.

b. are able to reduce waves energy so far better than coral reefs.

c. cannot protect coastlines good enough.

d. in some cases can decrease waves energy as same as coral reefs.

I. **Translate the following passage into your mother language. Write your translation in the space provided.**

Intertidal zones, also known as the foreshore or seashore, are the areas that are visible and exposed to air during low tide and covered up by saltwater during high tide. For the ecology study, intertidal region is an important model system because it contains a high variety of species and the zonation created by the tides causes species ranges to be compressed into very narrow bands. This situation led to easy study of species. The intertidal zone is divided into four physical divisions including the Spray zone, High intertidal zone, Middle Intertidal zone, and Low intertidal zone that each one having its distinct characteristics and wildlife. Intertidal zone organisms are adapted to tough extremes environments. Tides provide water regularly but it reneges from rain freshwater to highly saline water from drying between tidal inundations.

J. **Copy the technical terms and expressions used in this lesson. Then find your mother language equivalents of those terms and expressions and write them in the spaces provided.**

Technical term	Mother language equivalent
....	....
....	....
....	....
....	....
....	....
....	....
....	....
....	....
....	....
....	....

BIBLIOGRAPHY

Alexander, D.E. (1 May 1999). *Encyclopedia of Environmental Science*. Springer. ISBN 0-412-74050-8.

Brönmark, C., Hansson, L.A. (2005). *The Biology of Lakes and Ponds*. Oxford: Oxford University Press. p. 285. ISBN 0198516134.

Darmody, S.J. (1995). The Law of the Sea: A Delicate Balance for Environmental Lawyers. *Natural Resources & Environment* 9(4): 24–27. JSTOR 40923485.

Ellen MacArthur Foundation, McKinsey & Company, World Economic Forum (2016). *The New Plastics Economy—Rethinking the future of plastics (PDF)*. Ellen MacArthur Foundation. p. 29. Retrieved 11 January 2020.

Ferarrio, F., et al. (2014). The effectiveness of coral reefs for coastal hazard risk reduction and adaptation. *Nature Communications* 5:3794: 3794. Bibcode:2014NatCo...5.3794F. doi:10.1038/ncomms4794. PMC 4354160. PMID 24825660.

Focht, D.D. Biodegradation. *AccessScience*. doi:10.1036/1097-8542.422025.

Giller, S., Malmqvist, B. (1998). *The Biology of Streams and Rivers*. Oxford: Oxford University Press. p. 296. ISBN 0198549776.

Griffin, A. (1994). MARPOL 73/78 and Vessel Pollution: A Glass Half Full or Half Empty?. *Indiana Journal of Global Legal Studies* 1(2): 489–513. JSTOR 20644564.

Hildebrand, M., & Goslow, G.E. Jr. (2001). *Analysis of vertebrate structure. Principal ill. Viola Hildebrand*. New York: Wiley. p. 429. ISBN 0-471-29505-1.

http://marinespecies.org/introduced/wiki/Open_ocean_habitat

http://www.seasky.org/deep-sea/ocean-layers.html

https://byjus.com/biology/aquatic-ecosystem/

https://en.wikipedia.org/wiki/Abiotic_component

https://en.wikipedia.org/wiki/Aquatic_ecosystem

https://en.wikipedia.org/wiki/Coral_reef

https://en.wikipedia.org/wiki/Dilution_(equation)

https://en.wikipedia.org/wiki/Lake_ecosystem

https://en.wikipedia.org/wiki/Marine_habitats#Coastal

https://en.wikipedia.org/wiki/Marine_pollution

https://en.wikipedia.org/wiki/Pelagic_zone

https://en.wikipedia.org/wiki/Photodegradation

https://en.wikipedia.org/wiki/Primary_production

https://en.wikipedia.org/wiki/Wetland#Uses_of_wetlands

https://hokkaidowilds.org/water/kiritappu-wetland

https://marinebio.org/oceans/geography/

https://oceana.org/marine-life/marine-science-and-ecosystems/open-ocean

https://ourocean2018.org/

https://sciencing.com/characteristics-of-a-marine-biome-12535256.html

https://sciencing.com/definition-aquatic-ecosystem-6307480.html

https://sciencing.com/difference-between-temperate-tropical-ocean-8035716.html

https://theblogaquatic.org/aquatic-ecosystem-facts/

https://www.biologyonline.com/dictionary/saturated-soil

https://www.eligasht.com/Blog/wp-content/uploads/2018/06/01-6.jpg

https://www.nationalgeographic.com/environment/freshwater/aquatic-ecosystems/

https://www.nationalgeographic.org/encyclopedia/marine-pollution/

https://www.opodo.co.uk/blog/best-beaches-uk/

https://www.pollutionsolutions-online.com/news/waste-management/21/breaking-news/ what-is-plastic-photodegradation/35801

https://www.rainforestcruises.com/jungle-blog/amazon-river-facts

https://www.scientificamerican.com/article/oceans-are-warming-faster-than-predicted/

https://www.thegef.org/news/ocean-everyones-business

Keddy, P.A. (2010). *Wetland Ecology: Principles and Conservation* (2nd ed.). New York: Cambridge University Press. ISBN 978-0521519403.

Mann, K.H. 1973. Seaweeds: their productivity and strategy for growth. *Science* 182: 975–981.

Murray, N.J., Phinn, S.R., DeWitt, M., Ferrari, R., Johnston, R., Lyons, M.B., Clinton, N., Thau, D., & Fuller, R.A. (2019), The global distribution and trajectory of tidal flats. *Nature*, 565(7738): 222–225. Bibcode:2019Natur.565..222M, doi:10.1038/s41586-01 8-0805-8, PMID 30568300/

Physicalgeography.net Archived 26 January 2016 at the Wayback Machine. Physicalgeography.net. Retrieved on 29 December 2012.

Rowley, D.B. (2002). Rate of plate creation and destruction: 180 Ma to present. *Geological Society of America Bulletin* 114(8): 927–933. Bibcode:2002GSAB..114..927R. doi:1 0.1130/0016-7606(2002)114<0927:ROPCAD>2.0.CO;2.

Spalding, M., Ravilious, C., & Green, E. (2001). *World Atlas of Coral Reefs*. Berkeley, CA: University of California Press and UNEP/WCMC ISBN 0520232550.

United Nations (2017) Resolution adopted by the General Assembly on 6 July 2017, Work of the Statistical Commission pertaining to the 2030 Agenda for Sustainable Development (A/RES/71/313).

United States Environmental Protection Agency (2 March 2006). Marine Ecosystems. Retrieved 25 August 2006.

University of California Museum of Paleontology: The Marine Biome. Retrieved 27 September 2018.

US Department of Commerce, National Oceanic and Atmospheric Administration. "What is the intertidal zone?". oceanservice.noaa.gov. Retrieved 2019-03-21.

"What is the Intertidal Zone?". WorldAtlas. Retrieved 2019-09-17.

Where is Earth's water? Archived 14 December 2013 at the Wayback Machine, United States Geological Survey.

GLOSSARY OF TERMS

Abiotic component: Non-living chemical and physical components in the environment that have effect on living organisms and the functioning of ecosystems.

Biodegradation: Breakdown of organic matter by microorganisms such as bacteria and fungi.

Coral reefs: One of the most known underwater marine ecosystems that formed of coral skeletons from mostly intact coral colonies.

Dilution: The process of reducing the concentration of a solute in a solution which commonly done by the addition of solvent.

Kelp forests: Underwater areas with a high density of kelp, which is one of the most productive and dynamic ecosystems cover a vast part of the coastlines around the world.

Mudflats: Coastal wetlands that form in intertidal areas where tides or rivers caused the sediments accumulation.

Oceanic trenches: Topographic depressions in the seafloor that are very long, relatively narrow in width and the deepest parts of the ocean floor.

Poikilothermic animals: Animals whose internal temperature is variable considerably.

Photodegradation: Alteration of materials by photons, especially those found in the wavelengths of the sun's rays that is a very slow and typical process, refers to the combined action of sunlight and air.

Primary production: In ecology, primary production is the synthesis of organic compounds from atmospheric or aqueous carbon dioxide.

Saturated soil: Refers to a soil's water content when all voids (pores) spaces between soil particles are filled with water, either temporarily or permanently.

9 Water quality sciences

READING FOR COMPREHENSION

Water quality is the physical, chemical, and biological characteristics of water. It is a measure of the condition of water relative to the requirements of one or more biotic species and or to any human need or purpose. It is most frequently used by reference to a set of standards against which compliance can be assessed. The most common standards used to assess water quality relate to drinking water, safety of human contact, and for the health of ecosystems.

STANDARDS

In the setting of standards, agencies make political and technical/scientific decisions about how the water will be used. In the case of natural water bodies, they also make some reasonable estimate of pristine conditions. Different uses raise different concerns and therefore different standards are considered. Natural water bodies will vary in response to environmental conditions. Environmental scientists work to understand how these systems function which in turn helps to identify the sources and fates of contaminants. Environmental lawyers and policy makers work to define legislation that ensures that water is maintained at an appropriate quality for its identified use.

The vast majority of surface water on the planet is neither potable nor toxic. This remains true even if seawater in the oceans (which is too salty to drink) is not counted. Another general perception of *water quality* is that of a simple property that tells whether water is polluted or not. In fact, water quality is a very complex subject, in part because water is a complex medium intrinsically tied to the ecology of the Earth. Industrial pollution is a major cause of water pollution, as well as runoff from agricultural areas, urban stormwater runoff, and discharge of treated and untreated sewage (especially in developing countries).

CATEGORIES

The parameters for water quality are determined by the intended use. Work in the area of water quality tends to be focused on water that is treated for human consumption or in the environment.

HUMAN CONSUMPTION

Contaminants that may be in untreated water include microorganisms such as viruses and bacteria; inorganic contaminants such as salts and metals; organic

DOI: 10.1201/9781003293507-9

chemical contaminants from industrial processes and petroleum use; pesticides and herbicides; and radioactive contaminants. Water quality depends on the local geology and ecosystem, as well as human uses such as sewage dispersion, industrial pollution, use of water bodies as a heat sink, and overuse (which may lower the level of the water).

In the United States, the U.S. Environmental Protection Agency (EPA) limits the amounts of certain contaminants in tap water provided by public water systems. The Safe Drinking Water Act authorizes EPA to issue two types of standards: *primary standards* regulate substances that potentially affect human health, and *secondary standards* prescribe aesthetic qualities, those that affect taste, odor, or appearance. The U.S. Food and Drug Administration (FDA) regulations establish limits for contaminants in bottled water that must provide the same protection for public health. Drinking water, including bottled water, may reasonably be expected to contain at least small amounts of some contaminants. The presence of these contaminants does not necessarily indicate that the water poses a health risk.

Some people use water purification technology to remove contaminants from the municipal water supply they get in their homes, or from local pumps or bodies of water. For people who get water from a local stream, lake, or aquifer (well), their drinking water is not filtered by the local government.

ENVIRONMENTAL WATER QUALITY

Environmental water quality, also called ambient water quality, relates to water bodies such as lakes, rivers, and oceans. Water quality standards vary significantly due to different environmental conditions, ecosystems, and intended human uses. Toxic substances and high populations of certain microorganisms can present a health hazard for non-drinking purposes such as irrigation, swimming, fishing, rafting, boating, and industrial uses. These conditions may also affect wildlife that uses the water for drinking or as a habitat. Modern water quality laws generally specify protection of fisheries and recreational use and require as a minimum, retention of current quality standards (Figure 9.1).

There is some desire among the public to return water bodies to pristine or pre-industrial conditions. Most current environmental laws focus on the designation of uses. In some countries, these allow for some water contamination as long as the particular type of contamination is not harmful to the designated uses. Given the landscape changes in the watersheds of many freshwater bodies, returning to pristine conditions would be a significant challenge. In these cases, environmental scientists focus on achieving goals for maintaining healthy eco-systems and may concentrate on the protection of populations of endangered species and protecting human health.

MEASUREMENT

The complexity of water quality as a subject is reflected in the many types of measurements of water quality indicators. Some of the simple measurements listed below can be made on-site (temperature, pH, dissolved oxygen, conductivity, oxygen reduction potential (ORP), turbidity, Secchi Disk depth) in direct contact with the water source in

FIGURE 9.1   Water runoff, **Muscat, Oman** (By Lucy Ahmed).

question. More complex measurements that must be made in a lab setting require a water sample to be collected, preserved, and analyzed at another location. Making these complex measurements can be expensive. Because direct measurements of water quality can be expensive, ongoing monitoring programs are typically conducted by government agencies. However, there are local volunteer programs and resources available for some general assessment. Tools available to the general public are on-site test kits commonly used for home fish tanks and biological assessments.

Testing in Response to Natural Disasters and Other Emergencies

Inevitably after events such as earthquakes and Tsunamis, there is an immediate response by the aid agencies as relief operations get underway to try and restore basic infrastructure and provide the basic fundamental items that are necessary for survival and subsequent recovery. Access to clean drinking water and adequate sanitation is a priority at times like this. The threat of disease increases hugely due to the large numbers of people living close together, often in squalid conditions, and without proper sanitation.

After a natural disaster, as far as water quality testing is concerned there are widespread views on the best course of action to take and a variety of methods can be employed. The key basic water quality parameters that need to be addressed in an emergency are bacteriological indicators of fecal contamination, free chlorine residual, pH, turbidity, and possibly conductivity/TDS. There are a number of portable water test kits on the market widely used by aid and relief agencies for carrying out such testing.

The following is a list of indicators often measured by situational category:

Drinking Water

(Figure 9.2)

- Color of water
- pH
- Taste and odor (geosmin, 2-methylisoborneol (MIB), etc.)
- Dissolved metals and salts (sodium, chloride, potassium, calcium, manganese, magnesium)
- Microorganisms such as fecal coliform bacteria (*Escherichia coli*), Cryptosporidium, and Giardia lamblia
- Dissolved metals and metalloids (lead, mercury, arsenic, etc.)
- Dissolved organics: colored dissolved organic matter (CDOM), dissolved organic carbon (DOC)
- Radon
- Heavy metals
- Pharmaceuticals
- Hormone analogs

Environmental

Chemical assessment
- (also see salinity)
- Dissolved Oxygen (DO)
- nitrate-N
- orthophosphates
- Chemical oxygen demand (COD)
- Biochemical oxygen demand (BOD)
- Pesticides

(a)

(b)

FIGURE 9.2 Water sampling (a) from the channel (b) from the wetland (By Mohammad Albaji).

PHYSICAL ASSESSMENT

- pH
- Temperature
- Total suspended solids (TSS)
- Turbidity

BIOLOGICAL ASSESSMENT

Biological monitoring metrics have been developed in many places, and one widely used measure is the presence and abundance of members of the insect orders Ephemeroptera, Plecoptera, and Trichoptera. (Common names are, respectively, Mayfly, Stonefly, and Caddisfly.) EPT indexes will naturally vary from region to region, but generally, within a region, the greater the number of taxa from these orders, the better the water quality. EPA and other organizations in the United States offer guidance on developing a monitoring program and identifying members of these and other aquatic insect orders.

Individuals interested in monitoring water quality who cannot afford or manage lab scale analysis can also use biological indicators to get a general reading of water quality. One example is the IOWATER volunteer water monitoring program, which includes a benthic macroinvertebrate indicator key.

STANDARDS AND REPORTS

Canada

In Canada, Manitoulin Streams Improvement Association has become a leading model for water quality and fisheries rehabilitation. The association partners with landowners, farmers, fishermen, and the general public to improve water quality and the fisheries resource on Manitoulin Island and the Great Lakes. They do this by:

- Restricting livestock access to certain points on the river or installing alternative watering sources like nose pumps.
- Repair the Riparian Zone by planting trees and grasses to stabalize shorelines, provide habitat.
- Create in-stream habitat to increase fish and invertebrate populations.

Since 2000, Manitoulin Streams has rehabilitated 23 major sites on 4 waterways. They have had a Class Environmental Assessment conducted on 184 waterways on Manitoulin Island. The report identified 10 priority waterways that needed to be rehabilitated. Manitoulin Streams has conducted work on 4 of the 10 and has plans to work on a 5th, the Mindemoya River in the summer of 2010.

EUROPEAN UNION

The water policy of the European Union is primarily codified in three directives:

- The Urban Waste Water Treatment Directive (91/271/EEC) of 21 May 1991 concerning discharges of municipal and some industrial wastewaters;

- The Drinking Water Directive (98/83/EC) of 3 November 1998 concerning potable water quality;
- Water Framework Directive (2000/60/EC) of 23 October 2000 concerning water resources management.

UNITED KINGDOM

In England and Wales, acceptable levels for drinking water supply are listed in the Water Supply (Water Quality) Regulations 2000.

SOUTH AFRICA

Water quality guidelines for South Africa are grouped according to potential user types (e.g. domestic, industrial) in the 1996 Water Quality Guidelines. Drinking water quality is subject to the South African National Standard (SANS) 241 Drinking Water Specification.

UNITED STATES

In the United States, Water Quality Standards are created by state agencies for different types of water bodies and water body locations per desired uses. The Clean Water Act (CWA) requires each governing jurisdiction (states, territories, and covered tribal entities) to submit a set of biennial reports on the quality of water in their area. These reports are known as the 303(d), 305(b), and 314 reports, named for their respective CWA provisions, and are submitted to, and approved by, EPA. These reports are completed by the governing jurisdiction, typically a Department of Environmental Quality or similar state agency, and are available on the web. In the coming years, it is expected that the governing jurisdictions will submit all three reports as a single document, called the "Integrated Report". The 305(b) report (National Water Quality Inventory Report to Congress) is a general report on water quality, providing overall information about the number of miles of streams and rivers and their aggregate condition. The 314 report has provided similar information for lakes. The CWA requires states to adopt water quality standards for each of the possible designated uses that they assign to their waters. Should evidence suggest or document that a stream, river, or lake has failed to meet the water quality criteria for one or more of its designated uses, it is placed on the 303(d) list of impaired waters. Once a state has placed a water body on the 303(d) list, it must develop a management plan establishing Total Maximum Daily Loads for the pollutant(s) impairing the use of the water. These TMDLs establish the reductions needed to fully support the designated uses.

INTERNATIONAL STANDARDS

Water quality regulated by ISO is covered in the section of ICS 13.060, ranging from water sampling, drinking water, industrial class water, sewage water, and examination of water for chemical, physical, or biological properties. ICS 91.140.60 covers the standards of water supply systems.

EXERCISES

A. **Read each statement and decide whether it is true or false. Write "T" for true and "F" for false statements.**

TF1. The most common standards used to assess water quality relate to irrigation water safety for crops.

TF2. Oceans, seas, and lakes will vary in response to environmental conditions.

TF3. Environmental scientists work to define legislation that ensures that water is maintained at an appropriate quality for its identified use.

TF4. Industrial pollution is a major cause of water pollution.

TF5. Ambient water quality relates to water bodies such as lakes, rivers, and oceans.

B. **Circle a, b, c, or d which best completes the following items.**

1. Some people use .... technology to remove contaminants from the municipal water supply they get in their homes.
   a. water supply
   b. water purification
   c. water reuse
   d. wastewater

2. Water quality standards vary significantly due to different .... ecosystems and intended human uses.
   a. cultural conditions
   b. social conditions
   c. environmental conditions
   d. ecological conditions

3. The vast majority of .... on the planet is neither potable nor toxic.
   a. surface water
   b. subsurface water
   c. groundwater
   d. fossil water

4. Modern water quality laws generally specify protection of .... use and require as a minimum, retention of current quality standards.
   a. agricultural and household
   b. household and industrial
   c. fisheries and recreational
   d. agricultural and industrial

5. There are a number of .... test kits on the market widely used
   for water quality indicators tests.
   a. portable water
   b. wastewater
   c. polluted water
   d. contaminant water

C. **Match the sentence halves in Column I with their appropriate halves
   in Column II. Insert the letters a, b, c ... in the parentheses provided.
   There are more sentence halves in Column II than required.**

Column I	Column II
1. Water quality is	() **a.** is that of a simple property that tells water is whether polluted or not.
2. For people who get water from a local stream, lake, or well,	() **b.** is reflected in the many types of measurements of water quality indicators.
3. In the setting of standards,	() **c.** measure of the condition of water relative to the requirements of one or more biotic species.
4. The complexity of water quality as a subject	() **d.** their drinking water is not filtered by the local government.
5. Another general perception of water quality	() **e.** tends to be focused on water that is treated for human consumption or in the environment.
	() **f.** agencies make political and technical decisions about how the water will be used.
	() **g.** the physical, chemical, and biological characteristics of water.
	() **h.** they also make some reasonable estimate of pristine conditions.

D. **Cross out the word or words that make each statement false, and
   write the word or words that make each statement true in the blank.**
   1. Environmental lawyers and policy makers work to understand
      how these systems function which in turn helps to identify the
      sources and fates of contaminants.
      ....
   2. Microorganisms substances and high populations of certain
      Toxic can present a health hazard for non-drinking purposes.
      ....
   3. In fact, water quality is not a very complex subject, in part
      because water is not a complex medium intrinsically tied to the
      ecology of the Earth.
      ....
   4. Contaminants that may be in treated water include micro-
      organisms such as viruses and bacteria; inorganic contaminants
      such as salts and metals.
      ....

E. **Give answers to the following questions.**
  1. What is water quality science?
  2. What are the types of measurements of water quality indicators?
  3. What is the environmental water quality?

F. **For each word on the left, there are three meanings provided. Put a check mark (√) next to the choice which has the closest meaning to the word given.**

1. BOD	biochemical oxygen development	biochemical oxygen demand	biological oxygen demand
2. TSS	total suspended solids	total solids suspended	total salt solids
3. ORP	oxygen Reuse potential	oxygen Reduce potential	oxygen Reduction potential
4. TDS	total dissolved suspension	total dissolved solids	total dissolved salt
5. COD	chemical oxygen development	chemical oxygen demand	chemical oxygen decrease

G. **Fill in the blanks with the appropriate words from the following list.**

converting   dechlorination   concentrations   fish   toxicity
chlorine   harmful   chemically   bacteria   discharged

One of the most commonly used toxic compounds is ....
Chlorine is used throughout the world as a disinfectant to kill
.... organisms and to protect human health.
Unfortunately, the same characteristics that make it a good
disinfectant its ability to kill quickly in low .... also
make it harmful to the environment. Chlorine is used to kill harmful
organisms like .... and viruses at sewage treatment plants.
However, it may also kill or harm desirable organisms such as
.... and reptiles when the treated wastewater, called
effluent, is discharged back into the environment. Chlorine
.... can be avoided if the concentration of chlorine is
kept low and if the effluent is .... into a large, well-mixed
water body. Sometimes it is necessary to .... remove
chlorine from the effluent before discharging, to prevent toxicity.
This .... process is often accomplished by injecting sulfur

dioxide gas into the effluent. The sulfur dioxide gas reacts with the chlorine molecules, ... them into nontoxic chloride molecules.

H. **Read this passage and then circle a, b, c, or d which best completes the following items.**

The Federal Water Pollution Control Act, commonly known as the Clean Water Act (CWA), is the cornerstone of water quality legislation in the United States. It is not the result of a single piece of legislation. Rather, the current Act is a combination of federal water pollution control policies developed over many years. Its legislative history goes back over one hundred years to the Rivers and Harbors Act of 1899. The United States Congress brought together much of this historic water quality legislation in 1972 when they created Public Law 92–500, now simply called the Clean Water Act. The Clean Water Act consists of five separate parts called Titles. Title I is the introductory section, which declares the goals and policies of the Act. According to Title I:

The objective of this Act is to restore and maintain the chemical, physical, and biological integrity of the Nation's waters.

The main purpose of this passage is ....
a. water quality regulations.
b. water quality history.
c. the USA organizations for water quality.
d. protection of water quality.

I. **Translate the following passage into your mother language. Write your translation in the space provided.**

Since ancient times, villages have been built on riverbanks. Wastes from these villages were thrown into the rivers to be carried away. At first, few people lived downstream and the rivers had the natural capacity to assimilate the waste and cleanse themselves. This natural capacity for a water body to cleanse itself is called assimilative capacity. As the population continued to grow, however, the assimilative capacities of the waters were over-burdened and the rivers could no longer cleanse themselves. Today, most of us know it is unacceptable to discharge untreated waste into a river or stream. Waste dumped into a river upstream will be carried downstream to the users below. The phrase "we all live downstream" is often used to remind us to use our rivers wisely, respecting the rights of all downstream users. In turn, we hope the people living upstream from us will respect our rights. Although wastewater from most communities and industries is now routinely treated to remove pollutants, ultimately it is discharged into our rivers along with any pollutants that remain after treatment. Our efforts to keep rivers clean and healthy compete with this age-old practice of using our rivers to transport wastes.

Sometimes wastes enter our rivers and streams through more spread out, indirect, or diffuse discharges, or nonpoint source discharges. For instance, fertilizers, pesticides, and herbicides can be carried from our lawns and fields into nearby waters during and after rainstorms, as a result of stormwater runoff (Figure 9.3).

FIGURE 9.3   Karkeh Noor River, Hoveyzeh, Iran (By Mohammad Albaji).

J. **Copy the technical terms and expressions used in this lesson. Then find your mother language equivalents of those terms and expressions and write them in the spaces provided.**

Technical term	Mother language equivalent
....	....
....	....
....	....
....	....
....	....
....	....
....	....
....	....
....	....
....	....

BIBLIOGRAPHY

Clean Water Act, Section 303(d), 33 U.S.C. 1313; Section 305(b), 33 U.S.C. 1315(b); Section 314, 33 U.S.C. 1324.

Diersing, N. (2009). "Water Quality: Frequently Asked Questions". *PDA*. NOAA. http://floridakeys.noaa.gov/pdfs/wqfaq.pdf. Retrieved 2009-08-24.

Hodgson, K., & Manus, L. (2006). A drinking water quality framework for South Africa. *Water SA*. 2006 32(5): 673–678.

http://www.city-data.com/picfilesc/picc22826.php

http://www.hemayatonline.ir/detail/News/646

http://www.iso.org/iso/catalogue_ics_browse?ICS1=91&ICS2=140&ICS3=60&&published=
    on. Retrieved 29 February 2008.
http://www.nws.noaa.gov/om/hod/SHManual/SHMan014_glossary.htm
https://en.wikipedia.org/wiki/Water quality
International Organization for Standardization. "13.060: Water quality". http://www.iso.org/iso/
    iso_catalogue/catalogue_ics/catalogue_ics_browse.htm?ICS1=13&ICS2=060.   Retrieved
    29 February 2008.
IOWATER (Iowa Department of Natural Resources). Iowa City, IA. "Benthic Macro in-
    vertebrate Key".
Johnson, D.L., Ambrose, S.H., Bassett, T.J., Bowen, M.L., Crummey, D.E., Isaacson, J.S.,
    Johnson, D.N., Lamb, P., Saul, M., & Winter-Nelson, A.E. (1997). Meanings of en-
    vironmental terms. *Journal of Environmental Quality* 26: 581–589.
Kenneth, M., & Vigil, P.E. (2003). *Clean Water. An Introduction to Water Quality and Water
    Pollution Control*. Second Edition. Oregon State University Press Corvallis. ISBN
    0-87071-498-8. 181p.
Staley, C. & Pierson, G.S. (1899). *The Separate System of Sewerage, Its Theory and
    Construction*. New York: Van Nostrand.
United States Environmental Protection Agency (EPA). (2006). Washington, DC. "Water
    Quality Standards Review and Revision".
U.S. EPA. Washington, DC. "Methods for Measuring the Acute Toxicity of Effluents and
    Receiving Waters to Freshwater and Marine Organisms." Document No. EPA-821-R-
    02-012. October 2002.

GLOSSARY OF TERMS

Ambient: Refers to natural background conditions in the surrounding environment
    outside the zone in which water quality may be influenced by a discharge
    or source of contamination.

Contaminant: A substance that causes harm by contact or association, sewage or
    other materials that will render water unfit for its intended use, anything added
    to a substance that makes the substance impure or unfit for its intended use.

Drinking water: A water supply treated or untreated which is intended for human
    consumption and uses and which is considered to be free of toxins and patho-
    genic bacteria, cysts or viruses, potable water, fit to drink, potable water that has
    or is to be treated additionally, to enhance aesthetic quality and/or reduce mineral
    content plus other known or unknown, undesirable substances: by one or more
    point-of-use water processing devices or systems or purified bottled water.

Sewage: The waste and wastewater produced by residential and commercial
    sources and discharged into sewers.

Treated water: Water that has been subjected to a treatment process.

Water pollution: Degradation of a body of water by a substance or condition to
    such a degree that the water fails to meet specified standards or cannot be
    used for a specific purpose.

Waterbody: Waterbody is any significant accumulation of water, generally on a planet's
    surface. The term most often refers to oceans, seas, and lakes, but it includes
    smaller pools of water such as ponds, wetlands, or more rarely, puddles.

Water quality: The chemical, physical, and biological characteristics of water with
    respect to its suitability for a particular purpose.

10 Sewage treatment

READING FOR COMPREHENSION

Sewage treatment is the process of removing contaminants from municipal wastewater. It includes physical, chemical, and biological processes to remove contaminants and produce treated wastewater, called effluent, which is safe enough for release into the environment. A by-product of sewage treatment is sewage sludge that is suitable for disposal or reuse after being undergoing further treatment. The main use of sludge is in the agriculture due to its main constituent elements (carbon, nitrogen, and phosphorus) that create fertility potential (Figure 10.1).

ORIGINS OF SEWAGE

Residences and institutions

The residences wastewaters contain bodily wastes (primarily feces and urine), washing water, food preparation wastes, laundry wastes, and other waste products of normal living. This type of sewage is called domestic or sanitary sewage that enters into the sewer system through the toilets, baths, showers, kitchens, and sinks (Figure 10.2).

COMMERCIAL

Sewage of service establishments such as government offices, banks, shops, restaurants, called commercial sewage; however can be classified in the domestic sewage category if their characteristics are similar to household flows.

HOSPITAL

Hospitals wastewaters carrying damaging matters include pathogens, pharmaceuticals residues, chemical reagents, and radionuclide. Hospital wastewaters with hazardous substances should be treated according to regulatory status; however, some other hospital wastewaters have characteristics similar to domestic sewage.

INDUSTRIAL

Industrial sewage is the aqueous discard resulting from an industrial manufacturing process or the cleaning activities that take place along with that process. Industrial wastewater may contain some types of pollutants that cannot be removed by conventional sewage treatment.

FIGURE 10.1   Sewage treatment plant; (http://globalresearchenv.in/water-and-waste-water-management.html).

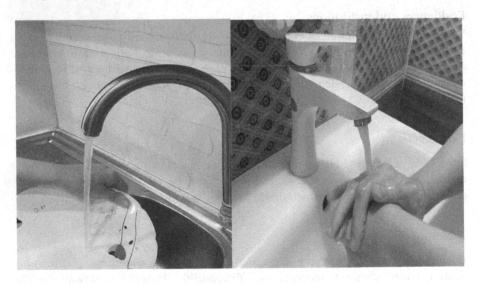

FIGURE 10.2   Domestic sewage (By Mohammad Albaji).

URBAN RUNOFF

Sewage may include urban runoff that is a fraction of precipitation. Impervious surfaces including roads, parking lots, and sidewalks built of materials such as asphalt and concrete, carry polluted stormwater to storm drains, instead of allowing the water to percolate through soil. This runoff is a major source of flooding and water pollution in urban area. Precipitation absorbs gases and particulates from the atmosphere, dissolves heavy metals, soil particles, organic compounds, animal

FIGURE 10.3 Urban runoff in Shahid Chamran University of Ahvaz, Iran (By Mohammad Albaji).

waste, washes spills, oil, grease, and debris from urban streets and highways and carries all these pollutants to a collection point (Figure 10.3).

STAGES OF THE SEWAGE TREATMENT PROCESS

Pretreatment

The pretreatment stage removes all large materials that are easy to separate from the **raw** wastewater. Items commonly removed during pretreatment include trash, tree limbs, cans, rags, sticks, plastic packets, etc. If these objects are not removed they damage or clog pumps and sewage lines of primary treatment clarifiers, and can cause considerable inefficiency in the process. These damages can be prevented by using a bar screen installed at the inlet of the sewage to the treatment plant. Bar screens or mesh screens come in varying designs of sizes to optimize solids removal. In modern plants, usually employ an automatic cleaning mechanism using electric motors and chains, while in smaller or less modern plants a cleaning manual plate may be used. The bar screen raises the water level by collecting objects;

FIGURE 10.4 Schematic drawing of a screen bar; (https://www.huber.de/huber-report/ ablage-bcrichte/screens/sturdy-screen-for-coarse-material-removal-huber-coarse-screen-trashmaxr.html).

therefore they must be removed regularly to prevent overflow. Separated objects from the sewage, termed screenings, are collected in dumpsters to be incinerated or disposed at landfills (Figure 10.4).

Grit Removal

Grit consists of sand, gravel, cinders, and other heavy solid materials that have higher specific gravity than the organic biodegradable solids in the sewage. Grit also consists organic particles in large (food waste) and small (eggshells, bone chips, seeds, and coffee grounds) sizes. Grit removal equipment is commonly located after screening and commination, and before primary treatment. Grit removal prevents unnecessary abrasion and wear of moving mechanical equipment, decreases the heavy deposits in aeration tanks, aerobic digesters, pipelines, channels, and conduits, and decreases the frequency of digester cleaning caused by grit excessive accumulations.

There are various types of grit removal systems:

- Aerated Grit Chamber
- Vortex-Type Grit Chamber
- Horizontal Flow Grit Chamber
- Hydrocyclone
- Detritus Tank

In order to select among these systems, the best balance must be established between the relevant considerations including the quantity and characteristics of grit, removal efficiency, potential adverse effects on downstream processes, head loss requirements, space requirements, organic content, and cost.

FLOW EQUALIZATION

Hydraulic surges are created by a variety of water-using appliances such as bathtubs, dishwashers, clothes washers, and shower facilities. These surges cause compromise and reduce efficiency in a sewage treatment system. The solution to this problem is flow equalization using equalization basins. Flow equalization prevents surges from forcing solids and organic material out of the treatment process, controls the flow in each stage of the treatment process, and provides adequate time to occur physical, biological, and chemical processes. Equalization basins are also used as a place for temporary storage of incoming sewage during plant maintenance and flow peaks, and as a means of diluting and distributing batch discharges of toxic or high-strength waste.

FAT AND GREASE REMOVAL

Many plants remove fat and grease using primary clarifiers with mechanical surface skimmers; however, in some larger plants the sewage flows through a small tank where skimmers collect the fat floating on the surface. In addition, air blowers, located at the bottom of the tank, may be used to help recover the fat as froth. Fat and grease from the floating material can be recovered for soap making.

PRIMARY TREATMENT

The primary treatment stage, also called the sedimentation stage, removes suspended solids from water using gravity. For this aim, the sewage is passed through large tanks commonly called pre-settling basins, primary sedimentation tanks, or primary clarifiers. The resulting sludge is then fed into a digester, in which further processing takes place. Primary treatment process removes about 60% of total suspended solids and about 35% of BOD but dissolved impurities are not removed (Figure 10.5).

SECONDARY TREATMENT

Secondary treatment is a part of the sewage treatment process that removes dissolved and colloidal compounds measured as biochemical oxygen demand (BOD). This stage usually applied to the liquid portion of sewage after primary treatment has removed settleable solids and floating material. Secondary treatment commonly uses indigenous, aquatic microorganisms in a managed aerobic habitat. Bacteria and protozoa consume biodegradable soluble organic contaminants while reproducing to form cells of biological solids. The processes of biological oxidation are sensitive to temperature factor and the rate of biological reactions increase with temperature.

FIGURE 10.5 Treatment clarifiers; (https://www.tpomag.com/online_exclusives/2017/04/optimize_your_clarifier_for_biological_phosphorus_removal_002y7).

Most surface aerated vessels occur in the temperature range of 4 °C to 32 °C. The United States Environmental Protection Agency (EPA) states that the effluent produced by the secondary treatment must contain a monthly average of less than 30 mg/l BOD and less than 30 mg/l suspended solids; however, weekly averages may be up to 50% higher.

There are two classes of secondary treatment systems; fixed-film or suspended-growth systems. Fixed-film systems, also called attached growth systems, are including trickling filters, constructed wetlands, bio-towers, and rotating biological contactors. In these systems, the biomass grows on media and the sewage passes over its surface. In the suspended-growth systems, including activated sludge, the biomass is mixed with the sewage. Fixed-film systems are more able to adapt to severe changes of the biological material amounts and can remove larger amounts of organic matter and suspended solids compare to suspended growth systems; however, the suspended-growth systems can be employed in a smaller space than fixed-film systems that treat the same amount of water (Figure 10.6).

TERTIARY TREATMENT

The target of tertiary treatment, also called effluent polishing, is more improvement of the effluent quality before it is discharged to the receiving environment (sea, river, lake, wetlands, ground, etc.). A treatment plant may use more than one tertiary treatment process. If disinfection is applied at this stage, it will be the last stage of the sewage treatment process.

FIGURE 10.6 Sewage treatment trickling filter beds; (https://www.dreamstime.com/photos-images/trickling-filter.html).

FILTRATION

Sand filtration removes much of the remained suspended matter, and the filtration over activated carbon, called carbon adsorption, causes removal of remaining toxins.

LAGOONS OR PONDS

More biological improvement of effluent may be achieved by storage in large ponds or lagoons constructed by humans. These lagoons are highly aerobic and colonization often is supported by native macrophytes, especially reeds. Small filter-feeding invertebrates such as Daphnia and species of Rotifera help in treatment through the removal of fine particulates.

BIOLOGICAL NUTRIENT REMOVAL

Sewage may contain large amounts of the nutrients nitrogen and phosphorus that excessive release of which to the environment can result in accumulation of nutrients. This process, called eutrophication, can contribute to the overgrowth of weeds, algae, and cyanobacteria (blue-green algae). Eutrophication may cause a rapid growth in the population of algae that are unsustainable and eventually most of them die. The algae decomposition by bacteria uses high levels of water oxygen and most or all of the animals die due to deoxygenating, which creates more organic matter for the bacteria to decompose. In addition, some algal species produce toxins that contaminate drinking water supplies.

There are different treatment processes for the removal of nitrogen and phosphorus. Nitrogen can be removed in the biological oxidation of nitrogen from ammonia to nitrate (nitrification), followed by the reduction of nitrate to nitrogen gas (denitrification). Nitrogen gas is removed from the water and released into the atmosphere. Phosphorus is removed through a biologically (enhanced biological phosphorus removal) or chemically (chemical precipitation, usually with salts of iron, aluminum, or lime) process.

DISINFECTION

Disinfection in the sewage treatment provides substantially decrease in the microorganisms number in the water to reuse in the drinking, bathing, irrigation, etc. The effective level of disinfection depends on the several factors such as quality of the water being treated (e.g., cloudiness, pH, etc.), the type of used disinfection, the disinfectant dosage (concentration and time), and other environmental variables. The common methods of disinfection are including ozone, chlorine, ultraviolet light, or sodium hypochlorite.

FOURTH TREATMENT STAGE

In the conventional treatment process (primary, secondary and tertiary treatment), micropollutants may not be removed and become a threat to the aquatic organisms although concentrations of those substances and their decomposition products are quite low. Micropollutants that can be mentioned are pharmaceuticals, ingredients of household chemicals, chemicals used in small businesses or industries, environmental persistent pharmaceutical pollutants (EPPP), or pesticides. The methods used for micropollutants removal mainly consist of activated carbon filters that adsorb the micropollutants.

ODOR CONTROL

Odors diffused by the sewage treatment process are commonly indicative of the anaerobic or septic condition. Primary stages of processing have tendency to create gases with foul smell in which hydrogen sulfide is the most complained. Large sewage treatment plants in urban areas for the elimination of the odors use techniques such as carbon reactors, a contact media with bio-slimes, small doses of chlorine, or circulating fluids to biologically capture and metabolize the noxious gases. Other methods of odor control exist, including addition of iron salts, hydrogen peroxide, calcium nitrate, etc. to manage hydrogen sulfide levels.

EXERCISES

A. **Read each statement and decide whether it is true or false. Write "T" for true and "F" for false statements.**

TF1. Sewage treatment includes physical, chemical, and biological processes to remove contaminants and produce treated wastewater.

TF2. Commercial sewage can be classified in the domestic sewage category if its characteristics be similar to household flows.

TF3. The pretreatment stage removes all large materials using treatment clarifiers.

TF4. Equalization basins are useful only to flow equalization.

TF5. Emitted odors by sewage treatment process indicate aerobic condition.

B. **Circle a, b, c, or d which best completes the following items.**

1. .... have high fertility potential due to its main elements (carbon, nitrogen, phosphorus) and mainly used in agriculture.
   a. Residences wastewaters
   b. Equalized flow
   c. Sludge
   d. Urban runoff

2. .... can cause considerable inefficiency, damage, and clog to the treatment plant equipment.
   a. Micropollutants
   b. Large objects
   c. Fat and grease
   d. Nutrients nitrogen and phosphorus

3. The frequency of digester cleaning can be decreased by ....
   a. grit removal.
   b. mesh screens.
   c. eutrophication.
   d. odor control.

4. Primary treatment processes are expected to remove about .... of total suspended solids and about ....…........... of BOD.
   a. 35%–60%
   b. 40%–65%
   c. 60%–35%
   d. 50%–50%

5. The methods of .... are mainly consist of activated carbon filters.
   a. odor control
   b. phosphorus removal
   c. flow equalization
   d. micropollutants removal

C. **Match the sentence halves in Column I with their appropriate halves in Column II. Insert the letters a, b, c ... in the parentheses provided. There are more sentence halves in Column II than required.**

Column I	Column II
1. Hydraulic surges	( ) **a.** is the removal of suspended solids from water by gravity.
2. Effluent	( ) **b.** may cause accumulation of nutrients
3. Water disinfection	( ) **c.** means controlling hydraulic velocity, or flow rate, through a wastewater treatment system.
4. Flow equalization	( ) **d.** controls the pathogenic microorganisms.
5. Sedimentation	( ) **e.** are the matters in wastewater, which will not stay in suspension during a preselected settling period and settles to the bottom.
	( ) **f.** created by sudden changes in the fluid velocity
	( ) **g.** is the reduction of nitrate to nitrogen gas
	( ) **h.** is the outflow of water or gas to a natural body of water

D. **Cross out the word or words that make each statement false, and write the word or words that make each statement true in the blank.**
   1. Commercial wastewaters contain damaging matters such as pathogens, pharmaceuticals residues, chemical reagents, and radionuclide.
      ....
   2. Secondary treatment removes large objects such as trash, tree limbs, cans, rags, sticks, plastic packets, etc.
      ....
   3. If eutrophication is applied at the tertiary treatment stage, it will be the last stage of the sewage treatment process.
      ....
   4. Small filter-feeding invertebrates such as Daphnia and species of Rotifera cause inefficiency in the treatment process by creating fine particulates.
      ....

E. **Give answers to the following questions.**
   1. What is the purpose of the sewage treatment process?
   2. What happens to separated objects from the sewage in pre-treatment stage?
   3. What are the types of grit removal systems?
   4. How do the high levels of nitrogen and phosphorus cause reduce in the water oxygen?

F. **For each word on the left, there are three meanings provided. Put a check mark (√) next to the choice which has the closest meaning to the word given.**

1. Domestic sewage	sanitary sewage	industrial sewage	effluent
2. Primary clarifiers	surface skimmers	primary sedimentation tanks	Aerated Grit Chamber
3. Fixed-film systems	grit removal systems	suspended-growth systems	attached growth systems
4. Tertiary treatment	effluent polishing	sedimentation stage	odor control stage
5. Septic condition	aerobic condition	anaerobic condition	high oxygen conditions

G. **Fill in the blanks with the appropriate words from the following list.**

BOD   activated   carbon   filters   urban   runoff   bar screens   flow   equalization   surface   skimmers   pre-settling basins   effluent   sludge

Sewage treatment purpose is to produce an .... that causes less damage to the receiving environment. There are different types of sewage based on the sewage source including domestic, commercial, hospital, industrial and .... Sewage treatment process may consist of these stages: pretreatment, primary treatment, secondary treatment, tertiary treatment, fourth treatment. The pretreatment stage removes all large materials using .... This stage may also consists ...., removal of grit, fat, and grease. The equipment that is used for these aims is equalization basins, grit removal systems, and .... In the primary treatment stage, suspended solids are separated from water by gravity. The sewage is passed through .... then the produced .... is fed into a digester. Secondary treatment removes dissolved and colloidal compounds measured as .... There are two main types of secondary treatment systems: fixed-film or suspended-growth systems. The tertiary treatment further improves the effluent quality before it is discharged to the environment. This stage consists of filtration, biological nutrient removal, and disinfection. The target of fourth treatment stage is the elimination of the micropollutants that may not be removed through a conventional treatment process (primary, secondary and tertiary treatment). The base of used methods for micropollutants removal is ....

H. **Read this passage and then circle a, b, c, or d which best completes the following items.**

In a typical sewage treatment plant, the cost of required energy usually accounts for approximately 30% of the annual operating costs. The amount of required energy depends on the treatment process type and wastewater load. For example, constructed wetlands require less energy than activated sludge plants because less energy is required for the aeration stage. Sewage treatment plants that produce biogas in their sewage sludge treatment process with anaerobic digestion can produce enough energy to provide most of the energy requirement by the sewage treatment plant itself. In commonly secondary treatment processes, most of the energy is spent for the dewatering and drying of sewage sludge by aeration, pumping systems, and equipment.

Some sewage treatment plants can provide their energy needs....

a. because less energy is required for the aeration stage.
b. by producing biogas through sewage sludge treatment process with anaerobic digestion.
c. due to low-load wastewater.
d. because 30% of the operating costs are paid in them.

I. **Translate the following passage into your mother language. Write your translation in the space provided.**

The accumulated sludge that is produced in a sewage treatment process must be treated and disposed of in a safe and effective way. The digestion aim is to decrease the amount of organic matter and the number of disease-causing microorganisms present in the solids. The more usual methods of sludge treatment include anaerobic digestion, aerobic digestion, composting, and incineration. Recently, the use of green approaches, such as phytoremediation, suggested as a valuable method to improve sewage sludge contaminated by trace elements and persistent organic pollutants. The selection of the treatment method depends on the amount of solids generated and other site-specific conditions. Small-scale plants typically employ composting while mid-scale plants use aerobic digestion, and larger-scale plants use anaerobic digestion.

J. **Copy the technical terms and expressions used in this lesson. Then find your mother language equivalents of those terms and expressions and write them in the spaces provided.**

Technical term	Mother language equivalent
....	....
....	....
....	....
....	....
....	....
....	....
....	....
....	....
....	....
....	....

BIBLIOGRAPHY

Astha Kumari, Nityanand Singh Maurya, Bhagyashree Tiwari, & Beychok, M.R. (1971). "Performance of surface-aerated basins". *Chemical Engineering Progress Symposium Series.* 67 (107): 322–339. Available at CSA Illumina website Archived 2007-11-14 at the Wayback Machine.

Burton, Jr., G.A. & Pitt, R.E. (2001). "Chapter 2. Receiving Water Uses, Impairments, and Sources of Stormwater Pollutants". *Stormwater Effects Handbook: A Toolbox for Watershed Managers, Scientists, and Engineers.* New York: CRC/Lewis Publishers. ISBN 978-0-87371-924-7.

DOE O 458.1 Chg 4 (LtdChg), Radiation Protection of the Public and the Environment.

EPA (1984). "Secondary Treatment Regulation: Special considerations." 40 CFR 133.103; and "Treatment equivalent to secondary treatment." 40 CFR 133.105.

EPA. Washington, DC (2004). "Primer for Municipal Waste water Treatment Systems." Document no. EPA 832-R-04-001.

Hammer, Mark J. (1975). *Water and Waste-Water Technology.* John Wiley & Sons ISBN 0-471-34726-4, pp. 223–225.

Hoffmann, H., Platzer, C., von Münch, E., & Winker, M. (2011). Technology review of constructed wetlands – Subsurface flow constructed wetlands for greywater and domestic wastewater treatment. Deutsche Gesellschaft für Internationale Zusammenarbeit (GIZ) GmbH, Eschborn, Germany, p. 11.

Hospital wastewater treatment scenario around the globe.

http://globalresearchenv.in/water-and-waste-water-management.html

http://www.kitchenbathguides.com/best-handheld-shower-head-ultimate-reviews

https://cotesautobody.com/5-important-reasons-wash-car-regularly

https://en.wikipedia.org/wiki/Aqueous_solution

https://en.wikipedia.org/wiki/Bar_screen

https://en.wikipedia.org/wiki/Clarifier

https://en.wikipedia.org/wiki/Effluent

https://en.wikipedia.org/wiki/Secondary_treatment
https://en.wikipedia.org/wiki/Sedimentation_(water_treatment)
https://en.wikipedia.org/wiki/Sewage
https://en.wikipedia.org/wiki/Sewage_treatment
https://en.wikipedia.org/wiki/Sewerage
https://en.wikipedia.org/wiki/Surge_control
https://en.wikipedia.org/wiki/Urban_runoff
https://www.advancedwaterinc.com/hand-washing-dishes-vs-dishwasher-comparison
https://www.americanrivers.org/threats-solutions/clean-water/stormwater-runoff/
https://www.britannica.com/technology/primary-treatment
https://www.crystaltanks.com/sewage_treatment_explained.html
https://www.directives.doe.gov/terms_definitions/settleable-solids
https://www.dreamstime.com/photos-images/trickling-filter.html
https://www.huber.de/huber-report/ablage-berichte/screens/sturdy-screen-for-coarse-
    material-removal-huber-coarse-screen-trashmaxr.html
https://www.initial.com/blog/washing-hands/
https://www.lenntech.com/processes/disinfection/what-is-water-disinfection.htm
https://www.norweco.com/learning-center/technical-resources/flow-equalization/
https://www.organicawater.com/primary-secondary-tertiary-wastewater-treatment-work/
https://www.sciencedirect.com/topics/earth-and-planetary-sciences/industrial-wastewater
https://www.suezwaterhandbook.com/water-and-generalities/what-water-should-we-treat-
    and-why/sludge/sludge-end-uses
https://www.tpomag.com/online_exclusives/2017/04/optimize_your_clarifier_for_biological_
    phosphorus_removal_002y7
https://www3.epa.gov/npdes/pubs/final_sgrit_removal.pdf
Huber Company, Berching, Germany (2012). "Sedimentation Tanks." Archived 2012-01-18
    at the Wayback Machine.
Margot, J. et al. (2013). Bacterial versus fungal laccase: potential for micropollutant de-
    gradation. *AMB Express*. 3 (1): 63. 10.1186/2191-0855-3-63. PMC 3819643. PMID
    24152339.
Nissim, W.G., Cincinelli, A., Martellini, T., Alvisi, L., Palm, E., Mancuso, S., & Azzarello, E. (July
    2018). Phytoremediation of sewage sludge contaminated by trace elements and organic
    compounds. *Environmental Research*. Elsevier. 164: 356–366. Bibcode:2018ER....164..356
    G. 10.1016/j.envres.2018.03.009. PMID 29567421. S2CID 5008369.
Omelia, C. (1998). Coagulation and sedimentation in lakes, reservoirs and water treatment
    plants. *Water Science and Technology* 37 (2): 129. 10.1016/S0273-1223(98)00018-3.
Ragsdale, F. Wastewater System Operators Manual (PDF). Ragsdale and Associates Training
    Specialists, LLC. Archived from the original (PDF) on 2013-03-19. Retrieved 2012-02-18.
"Runoff (surface water runoff)". USGS Water Science School. Reston, VA: U.S. Geological
    Survey (USGS). 2016-12-02.
Spellman, F.R. (2010). Spellman's Standard Handbook for Wastewater Operators, Volume 3.
    CRC Press. ISBN 978-1439818886.
Streicher, D. "Waste Water Treatment Plant Elmhurst, Illinois A Virtual Tour". Elmhurst
    College. Archived from the original on 29 March 2012. Retrieved 18 February 2012.
Tchobanoglous, G., Burton, F.L., Stensel, H.D., Metcalf & Eddy, Inc. (2003). *Wastewater
    Engineering: Treatment and Reuse* (4th ed.). McGraw-Hill. ISBN 978-0-07-112250-4.
UBA (Umweltbundesamt) (2014): Maßnahmen zur Verminderung des Eintrages von
    Mikroschadstoffen in die Gewässer. Texte 85/2014 (in German).
United States Environmental Protection Agency Wastewater Technology Fact Sheet
    Screening and Grit Removal.
Walker, J.D. and Welles Products Corporation (1976). "Tower for removing odors from
    gases." U.S. Patent No. 4421534.

Wastewater Engineering: Treatment and Resource Recovery. Tchobanoglous, G., Stensel, H. David, Tsuchihashi, R., Burton, F.L., Abu-Orf, M., & Bowden, G. (Fifth ed.). New York: McGraw-Hill. 2014. ISBN 978-0073401188. OCLC 858915999.

Water and Environmental Health at London and Loughborough (1999). "Waste water Treatment Options." Archived 2011-07-17 at the Wayback Machine Technical brief no. 64. London School of Hygiene & Tropical Medicine and Loughborough University.

Woodard & Curran, Inc., in Industrial Waste Treatment Handbook (Second Edition), 2006.

GLOSSARY OF TERMS

Aqueous solution: A solution in which the solvent is water.

Bar screen: A mechanical filter used to remove large objects, such as rags and plastics, from wastewater.

Water disinfection: The removal, deactivation, or killing of pathogenic microorganisms.

Effluent: An outflowing of water or gas to a natural body of water, from a structure such as a wastewater treatment plant, sewer pipe, or industrial outfall.

Flow equalization: The process of controlling hydraulic velocity, or flow rate, through a wastewater treatment system.

Hydraulic surges: Are created when the velocity of a fluid suddenly changes and becomes unsteady or transient.

Clarifiers: Settling tanks built with mechanical means for continuous removal of solids being deposited by sedimentation.

Sedimentation: A physical process in water treatment that uses gravity to remove suspended solids from water.

Settleable solids: That matter in wastewater, which will not stay in suspension during a preselected settling period, such as one hour, but settles to the bottom.

Sewer system: The infrastructure that conveys sewage or surface runoff using sewers. It is the system of pipes, chambers, manholes, etc. that conveys the sewage or stormwater.

11 Surface irrigation

READING FOR COMPREHENSION

Surface irrigation is defined as the group of application techniques where water is applied and distributed over the soil surface by gravity. It is by far the most common form of irrigation throughout the world and has been practiced in many areas virtually unchanged for thousands of years.

Surface irrigation is often referred to as flood irrigation, implying that the water distribution is uncontrolled and therefore, inherently inefficient. In reality, some of the irrigation practices grouped under this name involve a significant degree of management (for example surge irrigation). Surface irrigation comes in three major types; level basin, furrow, and border strip (Figure 11.1).

THE PROCESS

The process of surface irrigation can be described using four phases. As water is applied to the top end of the field it will flow or advance over the field length. The advance phase refers to that length of time as water is applied to the top end of the field and flows or advances over the field length. After the water reaches the end of the field it will either runoff or start to pond. The period of time between the end of the advance phase and the shut-off of the inflow is termed the wetting, ponding, or storage phase. As the inflow ceases the water will continue to runoff and infiltrate until the entire field is drained. The depletion phase is that short period of time after cut-off when the length of the field is still submerged. The recession phase describes the time period while the waterfront is retreating towards the downstream end of the field. The depth of water applied to any point in the field is a function of the opportunity time, the length of time for which water is present on the soil surface (Figures 11.2 and 11.3).

BASIN IRRIGATION

Level basin irrigation has historically been used in small areas having level surfaces that are surrounded by earth banks. The water is applied rapidly to the entire basin and is allowed to infiltrate. Basins may be linked sequentially so that drainage from one basin is diverted into the next once the desired soil water deficit is satisfied. A "closed" type basin is one where no water is drained from the basin. Basin irrigation is favored in soils with relatively low infiltration rates. Fields are typically set up to follow the natural contours of the land but the introduction of laser leveling and land grading has permitted the construction of large rectangular basins that are more appropriate for mechanized broadacre cropping. Basin irrigation is commonly used in the production of crops such as rice and wheat (Figure 11.4).

DOI: 10.1201/9781003293507-11

FIGURE 11.1   Flood irrigation in a citrus garden, Shahid Chamran University of Ahvaz, Iran (By Mohammad Albaji).

FIGURE 11.2   The advance phase. Shahid Chamran University of Ahvaz, Iran (By Mohammad Albaji).

FIGURE 11.3 The recession phase. Shahid Chamran University of Ahvaz, Iran (By Mohammad Albaji).

FURROW IRRIGATION

Furrow irrigation is conducted by creating small parallel channels along the field length in the direction of predominant slope. Water is applied to the top end of each furrow and flows down the field under the influence of gravity. Water may be supplied using gated pipe, siphon and head ditch, or bankless systems. The speed of water movement is determined by many factors such as slope, surface roughness, and furrow shape but most importantly by the inflow rate and soil infiltration rate. The spacing between adjacent furrows is governed by the crop species, common spacings typically range from 0.75 to 2 meters. The crop is planted on the ridge between furrows which may contain a single row of plants or several rows in the case of a bed type system. Furrows may range anywhere from less than 100 m to 2000 m long depending on the soil type, location, and crop type. Shorter furrows are commonly associated with higher uniformity of application but result in increased potential for runoff losses. Furrow irrigation is particularly suited to broad-acre row crops such as cotton, maize, and sugarcane. It is also practiced in various horticultural industries such as citrus, stone fruit, and tomatoes.

The water can take a considerable period of time to reach the other end, meaning water has been infiltrating for a longer period of time at the top end of the field. This

FIGURE 11.4   Level basin flood irrigation on wheat. Hoveyzeh, Iran (By Mohammad Albaji).

results in poor uniformity with high application at the top end with lower application at the bottom end. In most cases, the performance of furrow irrigation can be improved by increasing the speed at which water moves along the field (the advance rate). This can be achieved through increasing flow rates or through the practice of surge irrigation. Increasing the advance rate not only improves the uniformity but also reduces the total volume of water required to complete the irrigation (Figure 11.5).

Surge Irrigation

Surge Irrigation is a variant of furrow irrigation where the water supply is pulsed on and off in planned time periods (e.g. on for ½ hour off for ½ hour). The wetting and drying cycles reduce infiltration rates resulting in faster advance rates and higher uniformities than continuous flow. The reduction in infiltration is a result of surface consolidation, filling of cracks and micropores, and the disintegration of soil particles during rapid wetting and consequent surface sealing during each drying phase. The effectiveness of surge irrigation is soil type-dependent, for example, many clay soils experience a rapid sealing behavior under continuous flow therefore surge offers little benefit.

Bay/Border Strip Irrigation

Border strip or bay irrigation could be considered as a hybrid of level basin and furrow irrigation. The borders of the irrigated strip are longer and the strips are narrower than for basin irrigation and are orientated to align lengthwise with the slope

(a)

(b)

FIGURE 11.5 (a) Furrow irrigation system. Shahid Chamran University of Ahvaz, Iran. (b) Furrow irrigation system for sugarcane crop using gated pipes, Ahvaz, Iran (By Mohammad Albaji).

FIGURE 11.6 Border strip or bay irrigation. Hoveyzeh, Iran (By Mohammad Albaji).

of the field. The water is applied to the top end of the bay, which is usually constructed to facilitate free-flowing conditions at the downstream end. One common use of this technique includes the irrigation of pasture for dairy production (Figure 11.6).

ISSUES ASSOCIATED WITH SURFACE IRRIGATION

While surface irrigation can be practiced effectively using the right management under the right conditions, it is often associated with a number of issues undermining productivity and environmental sustainability:

- Waterlogging – Can cause the plant to shut down delaying further growth until sufficient water drains from the root zone. Waterlogging may be counteracted by drainage, tile drainage, or water table control by another form of subsurface drainage.
- Deep drainage – Over irrigation may cause water to move below the root zone resulting in rising water tables. In regions with naturally occurring saline soil layers (for example salinity in south-eastern Australia) or saline aquifers, these rising water tables may bring salt up into the root zone leading to problems of irrigation salinity.
- Salinization – Depending on water quality irrigation water may add significant volumes of salt to the soil profile. While this is a lesser issue for surface irrigation compared to other irrigation methods (due to the comparatively high leaching fraction), lack of subsurface drainage may restrict the leaching of

salts from the soil. This can be remedied by drainage and soil salinity control through flushing.

EXERCISES

A. **Read each statement and decide whether it is true or false. Write "T" for true and "F" for false statements.**

TF1. In surface irrigation, water distribution is uncontrolled and therefore, inherently efficient.

TF2. In the advance phase, water "advances" across the surface until the water extends over the entire area.

TF3. In Surge irrigation, water is applied rapidly to the entire basin and is allowed to infiltrate.

TF4. Border irrigation could be considered as a hybrid of level basin and furrow irrigation.

TF5. Basin irrigation is conducted by creating small parallel channels along the field length.

B. **Circle a, b, c, or d which best completes the following items.**

1. .... is the most common form of irrigation in the world.
   a. Border irrigation
   b. Furrow irrigation
   c. Surface irrigation
   d. Basin irrigation

2. The short period of time after cut-off when the length of the field is still submerged is ....
   a. storage phase.
   b. depletion phase.
   c. advance phase.
   d. recession phase.

3. The .... describes the time period while the waterfront is retreating toward the downstream end of the field.
   a. recession phase
   b. storage phase
   c. depletion phase
   d. advance phase

4. .... may be linked sequentially so that drainage from one is diverted into the next one.
   a. Border strips
   b. Bays
   c. Furrows
   d. Basins

5. .... can cause the plant to shut down delaying further growth until sufficient water drains from the root zone.
   a. Alkalinity
   b. Waterlogging
   c. Salinity
   d. Flooding

C. **Match the sentence halves in Column I with their appropriate halves in Column II. Insert the letters a, b, c ... in the parentheses provided. There are more sentence halves in Column II than required.**

Column I	Column II
1. Surface irrigation	() **a.** is suitable for light texture soils.
2. Opportunity time	() **b.** may result in rising water tables.
3. In furrow irrigation	() **c.** is a variant of furrow irrigation.
4. Surge irrigation	() **d.** may restrict the leaching of salts from the soil.
5. Over irrigation	() **e.** water may be supplied using gated pipe, siphon, and head ditch.
	f. includes three major types; level basin, furrow, and border strip.
	g. is the length of time for which water is present on the soil surface.
	h. refers to that length of time as water is applied to the top end of the field and flows or advances over the field length.

D. **Cross out the word or words that make each statement false, and write the word or words that make each statement true in the blank.**
   1. The period of time between the end of the advance phase and the shut-off of the inflow is termed the advance phase.

   ....

   2. Basin irrigation is favored in soils with relatively high infiltration rates.

   ....

   3. The spacing between adjacent furrows is governed by the inflow rate, common spacing typically ranges from 0.75 to 2 meters.

   ....

   4. In surge irrigation, the wetting and drying cycles increase infiltration rates resulting in faster advance rates and higher uniformities than continuous flow.

   ....

E. **Give answers to the following questions.**
   1. What are the most important factors that determined the speed of water in furrow irrigation?
   2. What are the issues with surface irrigation?
   3. What is the surge irrigation?

F. **For each word on the left, there are three meanings provided. Put a check mark (√) next to the choice which has the closest meaning to the word given.**

1. Salinity	waterlogging	drainage	salty
2. Irrigation	artificially watering	rainfall	precipitation
3. Bay irrigation	basin irrigation	surface irrigation	border irrigation
4. Storage phase	advance phase	recession phase	wetting phase
5. Recession phase	retreating phase	advance phase	depletion phase

G. **Fill in the blanks with the appropriate words from the following list.**
cut off   recession   interval   infiltrates   flow paths
ponding   depletion   decline   phases   advances

A surface irrigation event is composed of four .... When water is applied to the field, it .... across the surface until the water extends over the entire area. It may or may not directly wet the entire surface, but all of the .... have been completed. Then the irrigation water either runs off the field or begins to pond on its surface. The .... between the end of the advance and when the inflow is cut off is called the wetting or ....,.... ... phase. The volume of water on the surface begins to .... after the water is no longer being applied. It either drains from the surface (runoff) or .... into the soil. For the purposes of describing the hydraulics of the surface flows, the drainage period is segregated into the .... phase (vertical recession) and the .... phase (horizontal recession). Depletion is the interval between .... and the appearance of the first bare soil under the water. Recession begins at that point and continues until the surface is drained.

H. **Read this passage and then circle a, b, c, or d, which best completes the following items.**
Although surface irrigation is thousands of years old, the most significant advances have been made within the last decade. In the developed and industrialized countries, landholdings have become as much as 10–20 times as large, and the number of farm families has dropped sharply. Very large mechanized farming equipment has replaced animal-powered planting, cultivating, and harvesting operations. The precision of preparing the field for planting has improved by an order of

magnitude with the advent of the laser-controlled land grading equipment. Similarly, the irrigation works themselves are better constructed because of the application of high technology equipment.
In the lesser-developed countries ......

a. the number of farm families has dropped sharply.
b. landholdings have become as much as 10–20 times as large.
c. the very large farming equipment has not replaced animal-powered operations completely.
d. The precision of preparing the field for planting has very improved.

I. **Translate the following passage into your mother language. Write your translation in the space provided.**
Furrow irrigation avoids flooding the entire field surface by channeling the flow along the primary direction of the field using "furrows," "creases," or "corrugations." Water infiltrates through the wetted perimeter and spreads vertically and horizontally to refill the soil reservoir. Furrows are often employed in basins and borders to reduce the effects of topographical variation and crusting. There are several disadvantages with furrow irrigation. These may include: (1) an accumulation of salinity between furrows; (2) an increased level of tailwater losses; (3) the difficulty of moving farm equipment across the furrows; (4) the added expense and time to make extra tillage practice (furrow construction); (5) an increase in the erosive potential of the flow; (6) a higher commitment of labor to operate efficiently; and (7) generally furrow systems are more difficult to automate, particularly with regard to regulating an equal discharge in each furrow.

J. **Copy the technical terms and expressions used in this lesson. Then find your mother language equivalents of those terms and expressions and write them in the spaces provided.**

Technical term	Mother language equivalent
....	....
....	....
....	....
....	....
....	....
....	....
....	....
....	....
....	....
....	....

BIBLIOGRAPHY

Drainage Manual: A Guide to Integrating Plant, Soil, and Water Relationships for Drainage of Irrigated Lands. Interior Dept., Bureau of Reclamation. 1993. ISBN 0-16-061623-9.

Free articles and software on drainage of waterlogged land and soil salinity control. http://www.waterlog.info/. Retrieved 2010-07-28.

https://en.wikipedia.org/wiki/Surface_irrigation

http://today.cddn.co/related/Flood%20Irrigation

http://www.fao.org/docrep/r4082e/r4082e06.htm

http://www.fao.org/docrep/t0231e/t0231e04.htm

ILRI. (1989). Effectiveness and Social/Environmental Impacts of Irrigation Projects: a Review. In: Annual Report 1988, International Institute for Land Reclamation and Improvement (ILRI), Wageningen, the Netherlands, pp. 18–34.

Irrigation Glossary. (2015). Washington State University. http://irrigation.wsu.edu/Content/Resources/Irrigation-Glossary.php

Kemper, W.D., Trout, T.J., Humpherys, A.S., & Bullock, M.S. (1988). Mechanisms by which surge irrigation reduces furrow infiltration rates in a silty loam soil. *Transactions of the ASAE* 31(3): 921–829.

Walker, W.R. & Skogerboe, G.V. (1987). *Surface irrigation: Theory and practice.* Englewood Cliffs: Prentice-Hall.

GLOSSARY OF TERMS

Basin irrigation: A surface irrigation method where water is applied to a completely level area surrounded by dikes. On high clay content soils and managed correctly basin irrigation can be very efficient.

Border irrigation: Water is applied at the upper end of a strip having parallel berms or earthen dikes anywhere from 10 to 100 feet apart that serve to confine the water to a strip or rectangular area.

Flood irrigation: Also known as wild flood irrigation. Water is turned into a field without any flow control such as furrows, boarders, or corrugations. This is the least efficient, least uniform, and least effective method of irrigation.

Flow: The rate of water discharged from a source expressed in volume with respect to time.

Furrow irrigation: Also known as rill irrigation. Water is applied to row crops in small ditches or channels between the rows made by tillage implements.

Infiltration rate: Rate at which water can enter the soil. Usually measured in units of depth per unit time [in/hr, mm/hr].

Irrigation water: Water that is applied to assist crops in areas or during times where rainfall is inadequate.

Level basin: Water is applied at relatively high flow rates to a completely level field. Precision laser leveling is required to achieve the level fields suitable for this method.

Surface irrigation: Irrigation method where the soil surface is used to transport the water via gravity flow from the source to the plants. Common surface irrigation methods are furrow irrigation, corrugation irrigation, border irrigation, basin irrigation, and flood irrigation.

Surge irrigation: Water is sent down a furrow in pulses. After the initial surge, the furrow is allowed to dry. This wetting and drying cycle continues throughout the whole irrigation event. Surge increases uniformity for furrow irrigation and increases application efficiency.

12 Drip irrigation

READING FOR COMPREHENSION

Drip irrigation, also known as trickle irrigation or micro-irrigation, is an irrigation method that saves water and fertilizer by allowing water to drip slowly to the roots of plants, either onto the soil surface or directly onto the root zone, through a network of valves, pipes, tubing, and emitters (Figure 12.1).

HISTORY

Drip irrigation has been used since ancient times when buried clay pots were filled with water, which would gradually seep into the grass. Modern drip irrigation began its development in Afghanistan in 1866 when researchers began experimenting with irrigation using clay pipe to create combination irrigation and drainage systems. In 1913, E.B. House at Colorado State University succeeded in applying water to the root zone of plants without raising the water table. Perforated pipe was introduced in Germany in the 1920s and in 1934; O.E. Nobey experimented with irrigating through porous canvas hose at Michigan State University.

With the advent of modern plastics during and after World War II, major improvements in drip irrigation became possible. Plastic microtubing and various types of emitters began to be used in the greenhouses of Europe and the United States.

Usage of plastic emitter in drip irrigation was developed in Palestine by Simcha Blass and his son Yeshayahu. Instead of releasing water through tiny holes, blocked easily by tiny particles, water was released through larger and longer passageways by using velocity to slow water inside a plastic emitter. The first experimental system of this type was established in 1959 when Blass partnered with Kibbutz Hatzerim to create an irrigation company called Netafim. Together they developed and patented the first practical surface drip irrigation emitter. This method was very successful and subsequently spread to Australia, North America, and South America by the late 1960s.

In the United States, in the early 1960s, the first drip tape, called *Dew Hose*, was developed by Richard Chapin of Chapin Watermatics (first system established during 1964).

Beginning in 1989, Jain irrigation helped pioneer effective water management through drip irrigation in India. Jain irrigation also introduced the `Integrated System Approach', One-Stop-Shop for Farmers, and "Infrastructure Status to Drip Irrigation and Farm as Industry." The latest developments in the field involve even further reduction in drip rates being delivered and fewer tendencies to clog. In Pakistan, it has been promoted by the Pakistan Atomic Energy Commission, the Agriculture Development Bank as well as successive governments (Figure 12.2).

DOI: 10.1201/9781003293507-12

FIGURE 12.1 The emitters or drippers in action, Shahid Chamran University of Ahvaz, Iran (By Vahid Reza Pirozfar).

Modern drip irrigation has arguably become the world's most valued innovation in agriculture since the invention of the impact sprinkler in the 1930s, which offered the first practical alternative to surface irrigation. Drip irrigation may also use devices called micro-spray heads, which spray water in a small area, instead of dripping emitters. These are generally used on tree and vine crops with wider root zones. Subsurface drip irrigation (SDI) uses permanently or temporarily buried dripper line or drip tape located at or below the plant roots. It is becoming popular for row crop irrigation, especially in areas where water supplies are limited or recycled water is used for irrigation. Careful study of all the relevant factors like land topography, soil, water, crop, and agro-climatic conditions are needed to determine the most suitable drip irrigation system and components to be used in a specific installation.

FIGURE 12.2 Drip irrigation of corn field, Shush, Iran (By Mohammad Albaji).

COMPONENTS AND OPERATION

Components (listed in order from water source)

- Pump or pressurized water source
- Water Filter(s) – Filtration Systems: Sand Separator like Hydro-Cyclone, Screen filters, Media Filters, Automatic self-cleaning water filters
- Fertigation Systems (Venturi injector) and Chemigation Equipment (optional)
- Backwash Controller (Backflow Preventer)
- Pressure Control Valve (Pressure Regulator)
- Main Line (larger diameter Pipe and Pipe Fittings)
- Hand-operated, electronic, or hydraulic Control Valves and Safety Valves
- Smaller diameter polytube (often referred to as "laterals")
- Poly fittings and Accessories (to make connections)
- Emitting Devices at plants (ex. Emitter or Drippers, micro-spray heads, inline drippers, trickle rings)
- Note that in Drip irrigation systems Pumps and valves may be manually or automatically operated by a controller.

Most large drip irrigation systems employ some type of filter to prevent clogging of the small emitter flow path by small waterborne particles (Figure 12.3). New technologies are now being offered that minimize clogging. Some residential systems are installed without additional filters since potable water is already filtered at the water treatment plant. Virtually all drip irrigation equipment manufacturers recommend that filters be employed and generally will not honor warranty unless this is done. Last line filters just before the final delivery pipe are strongly recommended in addition to any other filtration system due to fine particle settlement and accidental insertion of particles in the intermediate lines.

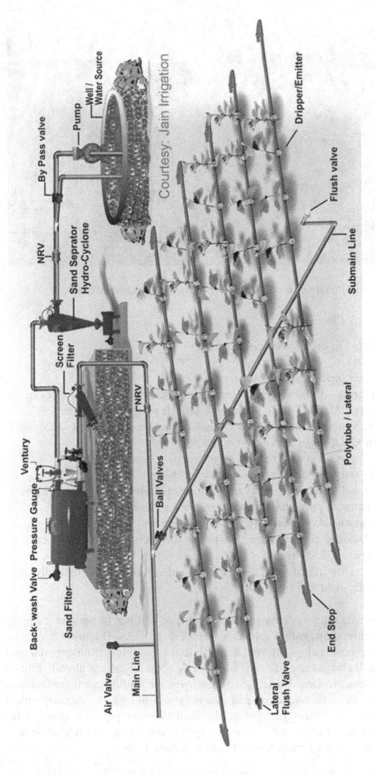

FIGURE 12.3 Drip irrigation system layout and its parts; (http://www.infonet-biovision.org/default/ct/293/soilconservation).

Drip and subsurface drip irrigation is used almost exclusively when using recycled municipal wastewater. Regulations typically do not permit spraying water through the air that has not been fully treated to potable water standards.

Because of the way the water is applied in a drip system, traditional surface applications of timed-release fertilizer are sometimes ineffective, so drip systems often mix liquid fertilizer with the irrigation water. This is called fertigation; fertigation and chemigation (application of pesticides and other chemicals to periodically clean out the system, such as chlorine or sulfuric acid) use chemical injectors such as diaphragm pumps, piston pumps, or venturi pumps. The chemicals may be added constantly whenever the system is irrigating or at intervals. Fertilizer savings of up to 95% are being reported from recent university field tests using drip fertigation and slow water delivery as compared to timed-release and irrigation by micro-spray heads.

If properly designed, installed, and managed, drip irrigation may help achieve water conservation by reducing evaporation and deep drainage when compared to other types of irrigation such as flood or overhead sprinklers since water can be more precisely applied to the plant roots. In addition, drip can eliminate many diseases that are spread through water contact with the foliage. Finally, in regions where water supplies are severely limited, there may be no actual water savings, but rather simply an increase in production while using the same amount of water as before. In very arid regions or on sandy soils, the preferred method is to apply the irrigation water as slowly as possible.

Pulsed irrigation is sometimes used to decrease the amount of water delivered to the plant at any one time, thus reducing runoff or deep percolation. Pulsed systems are typically expensive and require extensive maintenance. Therefore, the latest efforts by emitter manufacturers are focused toward developing new technologies that deliver irrigation water at ultra-low flow rates, i.e. less than 1.0 liter per hour. Slow and even delivery further improves water use efficiency without incurring the expense and complexity of pulsed delivery equipment.

Drip irrigation is used by farms, commercial greenhouses, and residential gardeners.

Drip irrigation is adopted extensively in areas of acute water scarcity and especially for crops such as coconuts, containerized landscape trees, grapes, bananas, ber, brinjal, citrus, strawberries, sugarcane, cotton, maize, and tomatoes (Figure 12.4).

ADVANTAGE AND DISADVANTAGES

The advantages of drip irrigation are:

- Minimized fertilizer/nutrient loss due to localized application and reduced leaching.
- High water application efficiency.
- Leveling of the field not necessary.
- Ability to irrigate irregular-shaped fields.
- Allows safe use of recycled water.
- Moisture within the root zone can be maintained at field capacity.
- Soil type plays less important role in frequency of irrigation.

FIGURE 12.4   Filtration systems for drip irrigation, Abadan, Iran (By Mohammad Albaji).

- Minimized soil erosion.
- Highly uniform distribution of water i.e. controlled by output of each nozzle.
- Lower labor cost.
- Variation in supply can be regulated by regulating the valves and drippers.
- Fertigation can easily be included with minimal waste of fertilizers.
- Foliage remains dry thus reducing the risk of disease.
- Usually operated at lower pressure than other types of pressurized irrigation, reducing energy costs.

THE DISADVANTAGES OF DRIP IRRIGATION ARE:

- Expense. Initial cost can be more than overhead systems.
- Waste. The sun can affect the tubes used for drip irrigation, shortening their usable life. Longevity is variable.
- Clogging. If the water is not properly filtered and the equipment is not properly maintained, it can result in clogging.
- Drip irrigation might be unsatisfactory if herbicides or top-dressed fertilizers need sprinkler irrigation for activation.
- Drip tape causes extra cleanup costs after harvest. You'll need to plan for drip tape winding, disposal, recycling, or reuse.
- Waste of water, time and harvest, if not installed properly. These systems require careful study of all the relevant factors like land topography, soil, water, crop, and agro-climatic conditions, and suitability of drip irrigation system and its components.
- Germination Problems. In lighter soils, subsurface drip may be unable to wet the soil surface for germination. Requires careful consideration of the installation depth.
- Salinity. Most drip systems are designed for high efficiency, meaning little or no leaching fraction. Without sufficient leaching, salts applied with the irrigation water may build up in the root zone, usually at the edge of the wetting pattern. On the other hand, drip irrigation avoids the high capillary potential of traditional surface-applied irrigation, which can draw salt deposits up from deposits below (Figure 12.5).

DRIPPER LINE

A dripper line is a type of drip irrigation tubing with emitters pre-installed at the factory.

EMITTER

An emitter is also called a dripper and is used to transfer water from a pipe or tube to the area that is to be irrigated. Typical emitter flow rates are from 0.16 to 4.0 US gallons per hour (0.6 to 16 L/h). In many emitters, flow will vary with pressure, while some emitters are *pressure compensating*. These emitters employ silicone diaphragms or other means to allow them to maintain a near-constant flow over a range of pressures, for example from 70 to 350 kPa.

FIGURE 12.5 Lemon tree and Olive tree with drip irrigation, Shahid Chamran University of Ahvaz, Iran (By Mohammad Albaji).

EXERCISES

A. **Read each statement and decide whether it is true or false. Write "T" for true and "F" for false statements.**

TF1. Micro-irrigation is a drip irrigation method that saves water and fertilizer.

TF2. Trickle irrigation use devices called micro-spray heads, which spray water in a small area, instead of dripping emitters.

TF3. Some large-scale drip irrigation systems are installed without filtration systems.

TF4. Sprinkler and drip irrigation are used, when using recycled municipal wastewater.

TF5. Drip irrigation help achieve water conservation by reducing evaporation and deep percolation.

B. **Circle a, b, c, or d which best completes the following items.**

1. Drip irrigation has been used since ancient times when buried .... were filled with water.

    a. tubes

    b. pipes

    c. clay pots

    d. clay pipe

2. In many emitters, flow will vary with pressure, while some emitters are ....
   a. pressure compensating.
   b. pressure controlling.
   c. pressure regulator.
   d. pressure-sensitive emitter.
3. In light textures soils .... may be unable to wet the soil surface for germination.
   a. drip irrigation
   b. subsurface drip irrigation
   c. c.surface irrigation
   d. sprinkler irrigation
4. In drip irrigation .... of the field is not necessary.
   a. fertility
   b. draining
   c. leaching
   d. leveling
5. Drip irrigation, usually operated at .... than sprinkler irrigation.
   a. higher pressure
   b. lower pressure
   c. the same pressure
   d. more pressure

C. **Match the sentence halves in Column I with their appropriate halves in Column II. Insert the letters a, b, c ... in the parentheses provided. There are more sentence halves in Column II than required.**

Column I	Column II
1. In drip irrigation systems pump and valves	( ) **a.** ability to irrigate irregular-shaped fields.
2. A dripper line	( ) **b.** use of recycled water.
3. An emitter	( ) **c.** employ silicone diaphragms.
4. In drip irrigation systems if the water	( ) **d.** is not properly filtered, it can result in clogging.
5. Drip irrigation can eliminate	( ) **e.** may be manually or automatically operated by a controller.
	( ) **f.** is a type of drip irrigation tubing with emitters pre-installed at the factory.
	( ) **g.** is also called a dripper and is used to transfer water from a pipe or tube to the irrigated area.
	( ) **h.** many diseases that are spread through water contact with the foliage.

D. **Cross out the word or words that make each statement false, and write the word or words that make each statement true in the blank.**
   1. Drip irrigation systems minimized nutrient loss due to localized application and increased leaching.

      ....
   2. In drip irrigation systems soil type plays more important role in frequency of irrigation.

      ....
   3. Nearly, all drip irrigation equipment manufacturers recommend that fertigation systems be employed and generally will not honor warranty unless this is done.

      ....
   4. In all drip irrigation systems, variation in supply can be regulated by regulating the pumps and drippers.

      ....

E. **Give answers to the following questions.**
   1. For what purpose are the filtration systems used?
   2. What are the advantages of drip irrigation systems?
   3. What are the disadvantages of drip irrigation systems?
   4. What are the components of drip irrigation systems?

F. **For each word on the left, there are three meanings provided. Put a check mark (√) next to the choice which has the closest meaning to the word given.**

1. Trickle	sprinkle	gravity	drip
2. Emitter	dropper	spray ring	dripper
3. Pressure Control Valve	pressure Regulator	venturi injector	up flow preventer
4. Pump	pressurized water source	water barrier	large machine
5. Backwash controller	up flow preventer	pressure control valve	backflow preventer

G. **Fill in the blanks with the appropriate words from the following list.**
   nozzles   larger plants   mainline   bubblers   emitters'   manufacturer
   micro-sprays   pressure-sensitive   products   pressure compensating

   There are two types of emitters: .... and pressure compensating. Pressure-sensitive emitters deliver a higher flow at higher water pressures. .... emitters provide the same flow over a wide pressure range. More .... made in recent years are pressure compensating. Turbulent flow and diaphragm emitters are non-plugging. Emitters can be attached into the .... or placed on the ends of ¼ inch microtubes. Because .... are generally color-coded by flow rates, purchase all emitters from one .... because color codes differ among manufacturers.

.... are devices that emit higher flows of water in a circular pattern. They are useful for irrigating .... such as roses and shrubs and for filling basins around newly planted trees or shrubs. Some can be adjusted for flows from 0 to 35 gph.

.... emit large droplets or fine streams of water just above the ground. They are available with .... in full, half, and quarter circle patterns that wet diameters varying from 18 inches to 12 feet. They should be placed on a separate zone from other drip devices because of their greater water use that can vary from 7 to 25 gph.

H. **Read this passage and then circle a, b, c, or d which best completes the following items.**

Drip, or micro-irrigation, technology uses a network of plastic pipes to carry a low flow of water under low pressure to plants. Water is applied much more slowly than with sprinkler irrigation. Drip irrigation exceeds 90% efficiency whereas sprinkler systems are 50% to 70% efficient. It is so efficient that many water utilities exempt landscapes irrigated with drip from restrictions during drought. Note that any irrigation system is only as efficient as the watering schedule used. If systems are set to water excessively, any system including drip can wastewater.

Drip irrigation is said to be better than sprinkler irrigation because .....

  a. it is useful in drought.

  b. it uses a network of plastic pipes to irrigate the land.

  c. the water is applied much more slowly than with sprinkler irrigation.

  d. it is more efficient than sprinkler irrigation systems.

I. **Translate the following passage into your mother language. Write your translation in the space provided.**

Drip irrigation delivers water slowly immediately above, on, or below the surface of the soil. This minimizes water loss due to runoff, wind, and evaporation. Drip irrigation can be operated during the windy periods frequently seen in Colorado. Drip systems can be managed with an AC or battery-powered controller. Automated landscape irrigation is an advantage to many people with busy lifestyles. Adaptable and changeable over time, drip systems can be easily expanded to irrigate additional plants if water is available. Emitters can be simply exchanged or removed and emitter lines eliminated or repositioned. When plants are removed or die, drip lines should be plugged.

If emitters are poorly placed, too far apart, or too few in number, root development may be restricted by the limited soil area wetted. Water seeping at ground level is hard to see and makes it difficult to know if the system is working properly. An indicator device that raises and lowers a flag to show when water is flowing is available to overcome this issue. Regular maintenance inspections are needed to maintain system effectiveness – the same as with high-pressure sprinkler systems. Clogs are much less likely with filtered water and proper pressure regulation used in combination with self-cleaning emitters.

J. **Copy the technical terms and expressions used in this lesson. Then find your mother language equivalents of those terms and expressions and write them in the spaces provided.**

Technical term	Mother language equivalent
....	....
....	....
....	....
....	....
....	....
....	....
....	....
....	....
....	....
....	....

BIBLIOGRAPHY

Drip and Micro Irrigation Design and Management for Trees, Vines, and Field Crops, 3rd Edition, by Charles M. Burt and Stuart W. Styles, published by the Irrigation Training and Research Center, 2007.
https://en.wikipedia.org/wiki/Drip_irrigation
https://en.wikipedia.org/wiki/Jalgaon
http://www.davidfrisk.com/blog/2014/7/30/in-my-garden-my-new-drip-irrigation-system.html
http://www.infonet-biovision.org/default/ct/293/soilconservation
Irrigation, 5th Edition,Engr Muhammad Irfan Khan Yousafzai, Claude H. Pair, editor, published by the Irrigation Association, 1983.
Maintenance Manual, published by Jain Irrigation, 1989.
Metcalf, L. & Harrison, P. (1922). *Sewerage and Sewage Disposal: A Textbook*. New York: McGraw-Hill.
Nakayama, F.S. & Bucks, D.A. (1986). Trickle Irrigation for Crop Production, published by Elsevier, 1986, ISBN 0-444-42615-9.
Wilson, C. & Bauer, M. (2014). Drip Irrigation for Home Gardens. Colorado State University. Fact Sheet No. 4.702. Gardening Series, Basics. www.ext.colostate.edu.

GLOSSARY OF TERMS

Backflow prevention device: A device that prevents contaminated water from being sucked back into the water source should a reverse flow situation occur.

Bubbler: A water emission device that tends to bubble water directly onto the ground or that throws water a short distance.

Chemigation: The process of applying chemicals (fertilizers, insecticides, herbicides, etc.) to crops or soil through an irrigation system with the water.

Control valve: A device used to control the flow of water.

Drip irrigation: Also known as trickle or micro-irrigation. Water is applied at very low flow rates (drip) through emitters directly to the soil. Emitter flow rates are generally less than 3 gallons per hour.

Emitter: Also known as a dripper. Used in drip irrigation to control the flow of water into the soil.

Fertigation: Application of fertilizers through the irrigation system. A form of chemigation.

Mainline: Tubing used in the drip system and is sometimes called lateral line.

Micro-irrigation: An updated term adopted by the American Society of Agricultural Engineers for drip or trickle irrigation that also includes micro-spray and other new devices operating at low pressures. Water is applied frequently just above, on, or below the surface of the soil at low flow rates with the goal of placing a quantity of water in the root zone that nearly approximates the consumptive use of the plant.

Micro-spray: A low-pressure sprayer device generally placed on a stake that is designed to wet soil with a fan or jet of water.

Pressure-compensating emitter: An emitter is designed to maintain a constant output (flow) over a wide range of operating pressures and elevations. Pressure-sensitive emitter – an emitter.

13 Sprinkler irrigation

READING FOR COMPREHENSION

In sprinkler or overhead irrigation, water is piped to one or more central locations within the field and distributed by overhead high-pressure sprinklers or guns. A system utilizing sprinklers, sprays, or guns mounted overhead on permanently in-stalled risers is often referred to as a *solid-set* irrigation system. Higher pressure sprinklers that rotate are called *rotors* and are driven by a ball drive, gear drive, or impact mechanism. Rotors can be designed to rotate in a full or partial circle. Guns are similar to rotors, except that they generally operate at very high pressures of 40 to 130 lbf/in² (275 to 900 kPa) and flows of 50 to 1200 US gal/min (3 to 76 L/s), usually with nozzle diameters in the range of 0.5 to 1.9 inches (10 to 50 mm). Guns are used not only for irrigation but also for industrial applications such as dust suppression and logging (Figure 13.1).

Sprinklers can also be mounted on moving platforms connected to the water source by a hose. Automatically moving wheeled systems known as *traveling sprinklers* may irrigate areas such as small farms, sports fields, parks, pastures, and cemeteries unattended. Most of these utilize a length of polyethylene tubing wound on a steel drum. As the tubing is wound on the drum powered by the irrigation water or a small gas engine, the sprinkler is pulled across the field. When the sprinkler arrives back at the reel the system shuts off. This type of system is known to most people as a "water reel" traveling irrigation sprinkler and they are used extensively for dust suppression, irrigation, and land application of wastewater. Other travelers use a flat rubber hose that is dragged along behind while the sprinkler platform is pulled by a cable. These cable-type travelers are definitely old technology and their use is limited in today's modern irrigation projects (Figure 13.2).

CENTER PIVOT

Center pivot irrigation is a form of sprinkler irrigation consisting of several seg-ments of pipe (usually galvanized steel or aluminum) joined together and supported by trusses, mounted on wheeled towers with sprinklers positioned along its length. The system moves in a circular pattern and is fed with water from the pivot point at the center of the arc. These systems are found and used in all parts of the world and allow irrigation of all types of terrain. Newer systems have drop sprinkler heads as shown in the image that follows.

Most center pivot systems now have drops hanging from a u-shaped pipe at-tached at the top of the pipe with sprinkler heads that are positioned a few feet (at most) above the crop, thus limiting evaporative losses. Drops can also be used with drag hoses or bubblers that deposit the water directly on the ground between crops. Crops are often planted in a circle to conform to the center pivot. This type of

DOI: 10.1201/9781003293507-13

FIGURE 13.1   Sprinkler irrigation of a garden, Shush, Iran (By Mohammad Albaji).

system is known as LEPA (Low Energy Precision Application). Originally, most center pivots were water powered. These were replaced by hydraulic systems (*T-L Irrigation*) and electric motor-driven systems (Reinke, Valley, Zimmatic). Many modern pivots feature GPS devices (Figures 13.3 and 13.4).

LATERAL MOVE (SIDE ROLL, WHEEL LINE)

A series of pipes, each with a wheel of about 1.5 m diameter permanently affixed to its midpoint and sprinklers along its length, are coupled together at one edge of a field. Water is supplied at one end using a large hose. After sufficient water has been applied, the hose is removed and the remaining assembly rotated either by hand or with a purpose-built mechanism so that the sprinklers move 10 m across the field. The hose is reconnected. The process is repeated until the opposite edge of the field is reached. This system is less expensive to install than a center pivot but much more labor intensive to operate, and it is limited in the amount of water it can carry. Most systems utilize 4 or 5-inch (130 mm) diameter aluminum pipe. One feature of a lateral move system is that it consists of sections that can be easily disconnected. They are most often used for small or oddly a shaped field such as those found in hilly or mountainous regions, or in regions where labor is inexpensive (Figure 13.5).

FIGURE 13.2 A traveling sprinklers, Shahid Chamran University of Ahvaz, Iran (By Mohammad Albaji).

ADVANTAGES OF SPRINKLER SYSTEMS

a. Sprinkler irrigation does not require surface shaping or leveling.
b. Can be applied to areas of variable topography.
c. Suitable for most crops, not all, and are adaptable to most irrigable soils.
d. Flexibility is possible because sprinkler heads are available in a wide range of discharge capacities.
e. Chemical and fertilizer applications are easily used with sprinkler systems.

DISADVANTAGES OF SPRINKLER SYSTEMS

a. Water application efficiency under sprinkler irrigation is strongly affected by wind.
b. Some crops are particularly sensitive and may suffer leaf scorch because of the salts deposited on the leaves as the intercepted irrigation water evaporates.
c. Some crops are especially sensitive to fungal diseases, leaf scorch, or fruit damage, and tall crops may obstruct hand-move or side-roll portable systems.

FIGURE 13.3   A center pivot system; (http://tlirr.com/products/center-pivot/#prettyPhoto/1/).

FIGURE 13.4   LEPA (Low Energy Precision Application) system. (https://www.senninger.com/site-study/saving-water-increasing-alfalfa-yield-lepa-close-spacing).

    d. Falling drops on bare soil cause slaking and surface sealing (crusting) which can be severe when the sodium ion predominates in the water affecting the soil's clay fraction.

    e. High maintenance requirements, constant and meticulous maintenance of sprinkle irrigation systems is crucial if these systems are to justify their costs.

FIGURE 13.5 Wheel line irrigation system, Hamedan, Iran (By Mehdi Jovzi).

f. High operating pressures.

g. The danger of system failure increases with technological complexity and requirements of expertise and quick availability of spare parts.

h. A malfunction of one of numerous parts can soon transform a working marvel of technology into a standing monument of inefficiency.

EXERCISES

A. **Read each statement and decide whether it is true or false. Write "T" for true and "F" for false statements.**

TF1. In sprinkler irrigation, water is distributed by emitters.

TF2. Traveling sprinklers move automatically by wheels.

TF3. Center pivot irrigation moves in a circular pattern and is fed with water from the pivot point.

TF4. In the Wheel line irrigation system, water is supplied at the center using a large hose.

TF5. Sprinkler irrigation systems can be applied to areas of different topography.

B. **Circle a, b, c, or d which best completes the following items.**

1. Sprinkler irrigation systems that use permanently installed risers are known as ....

   a. center pivot irrigation system.

   b. solid-set irrigation system.

   c. wheel line irrigation system.

   d. linear irrigation system.

2. .... is a form of sprinkler irrigation consisting of several segments of steel pipe joined together and supported by trusses, mounted on wheeled towers.
   a. Center pivot irrigation system
   b. Wheel line irrigation system
   c. Solid-set irrigation system.
   d. Gun irrigation system.
3. .... system is less expensive to install than .... system, but much more labor intensive to operate, and it is limited in the amount of water it can carry.
   a. A wheel line – a lateral move
   b. A center pivot – a wheel line
   c. A wheel line – a center pivot
   d. A wheel line – a side roll
4. Sprinkler irrigation systems do not require ....
   a. leveling.
   b. leaching.
   c. draining.
   d. wind breaking.
5. Water application efficiency under sprinkler irrigation is strongly affected by ....
   a. water salinity.
   b. wind.
   c. water alkalinity.
   d. water acidity.

C. **Match the sentence halves in Column I with their appropriate halves in Column II. Insert the letters a, b, c ... in the parentheses provided. There are more sentence halves in Column II than required.**

Column I	Column II
1. Most center pivot systems	() **a.** for dust suppression and logging.
2. One feature of a lateral move system	() **b.** is crucial if these systems are to justify their costs.
3. Guns are used not only for irrigation but also	() **c.** causing slaking and surface sealing (crusting).
4. Chemical and fertilizer applications	() **d.** now have drops hanging from a u-shaped pipe.
5. Constant and meticulous maintenance	() **e.** are high operating pressures of sprinkle irrigation systems.
	() **f.** are easily used with sprinkler systems.
	() **g.** falling drops on bare soil.
	() **h.** is that it consists of sections that can be easily disconnected.

D. **Cross out the word or words that make each statement false, and write the word or words that make each statement true in the blank.**

   1. Rotors are similar to guns rotors, except that they generally operate at very high pressures of 40 to 130 lbf/in² (275 to 900 kPa) and flows of 50 to 1200 US gal/min (3 to 76 L/s), usually with nozzle diameters in the range of 0.5 to 1.9 inches (10 to 50 mm).

      ....

   2. Center pivot irrigation systems are found and used in all parts of Asia and allow irrigation of all types of terrain.

      ....

   3. In sprinkler systems, flexibility is possible because sprinkler heads are available in a wide range of discharge water capacities.

      ....

   4. In sprinkler irrigation systems, the danger of system failure decreases increases with technological complexity and requirements of expertise and quick availability of spare parts.

      ....

E. **Give answers to the following questions.**

   1. For what purpose the falling drops on bare soil is not useful?
   2. What are the advantages of sprinkler irrigation systems?
   3. What are the disadvantages of sprinkler irrigation systems?
   4. What are the general components of center pivot irrigation system?

F. **For each word on the left, there are three meanings provided. Put a check mark (√) next to the choice which has the closest meaning to the word given.**

1. Lateral move	gun irrigation	center pivot irrigation	wheel line
2. Sprinkler	overhead	under head	drip
3. Traveling sprinkler irrigation	gun irrigation	solid-set irrigation	water reel irrigation
4. Pressurized irrigation	surface irrigation	drip irrigation	subsurface irrigation
5. LEPA	low energy program application	low energy precision agriculture	low energy precision application

G. **Fill in the blanks with the appropriate words from the following list.**

runoff pumping pressure application land contour sprinkler irrigation natural rainfall infiltration rates sprinklers ground adaptable recommended

Sprinkler irrigation is a method of applying irrigation water which is similar to .... Water is distributed through a system of pipes usually by .... It is then sprayed into the air through sprinklers so that it breaks up into small water drops which fall to the .... The pump supply system, sprinklers, and operating conditions must be designed to enable a uniform .... of water.

.... is suited for most row, field and tree crops and water can be sprayed over or under the crop canopy. However, large sprinklers are not .... for irrigation of delicate crops such as lettuce because the large water drops produced by the .... may damage the crop.

Sprinkler irrigation is .... to any farmable slope, whether uniform or undulating. The lateral pipes supplying water to the sprinklers should always be laid out along the .... whenever possible. This will minimize the .... changes at the sprinklers and provide a uniform irrigation.

Sprinklers are best suited to sandy soils with high .... although they are adaptable to most soils. The average application rate from the sprinklers (in mm/hour) is always chosen to be less than the basic infiltration rate of the soil, so that surface ponding and .... can be avoided.

H. **Read this passage and then circle a, b, c, or d which best completes the following items.**

As water sprays from a sprinkler it breaks up into small drops between 0.5 and 4.0 mm in size. The small drops fall close to the sprinkler whereas the larger ones fall close to the edge of the wetted circle. Large drops can damage delicate crops and soils and so in such conditions it is best to use the smaller sprinklers. Drop size is also controlled by pressure and nozzle size. When the pressure is low, drops tend to be much larger as the water jet does not break up easily. So to avoid crop and soil damage use small diameter nozzles operating at or above the normal recommended operating pressure.

The main purpose of this passage is ....

a. sprinkler application rate
b. sprinkler drop sizes
c. sprinkler wetting patterns
d. sprinkler pressure rate

I. **Translate the following passage into your mother language. Write your translation in the space provided.**

The mainline – and sub mainlines – are pipes that deliver water from the pump to the laterals. In some cases, these pipelines are permanent and are laid on the soil surface or buried below ground. In other cases, they are temporary and can be moved from field to field. The main pipe materials used include asbestos cement, plastic, or aluminum alloy.

The laterals deliver water from the mainline or sub mainlines to the sprinklers. They can be permanent but more often they are portable and made of aluminum alloy or plastic so that they can be moved easily.

The most common type of sprinkler system layout is shown in the followed figure. It consists of a system of lightweight aluminum or plastic pipes which are moved by hand. The rotary sprinklers are usually spaced 9–24 m apart along the lateral which is normally 5–12.5 cm in diameter. This is so it can be carried easily. The lateral pipe is located in the field until the irrigation is complete. The pump is then switched off and the lateral is disconnected from the mainline and moved to the next location. It is re-assembled and connected to the mainline and the irrigation begins again. The lateral can be moved one to four times a day. It is gradually moved around the field until the whole field is irrigated. This is the simplest of all systems. Some use more than one lateral to irrigate larger areas (Figure 13.6).

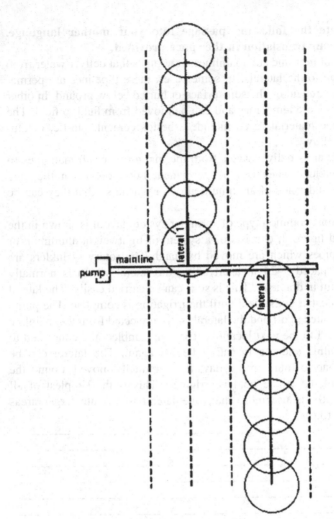

FIGURE 13.6 Hand-moved sprinkler system using two laterals (Laterals 1 and 2 in position 2); (http://www.fao.org/docrep/s8684e/s8684e06.htm).

J. **Copy the technical terms and expressions used in this lesson. Then find your mother language equivalents of those terms and expressions and write them in the spaces provided.**

Technical term	Mother language equivalent
....	....
....	....
....	....
....	....
....	....
....	....
....	....
....	....
....	....
....	....

BIBLIOGRAPHY

Davis, M. (2006). *City of Quartz: Excavating the Future in Los Angeles*. London: Verso. p. 233. ISBN 978-1-84467-568-5.

https://en.wikipedia.org/wiki/Sprinkler_irrigation

https://en.wikipedia.org/wiki/Water_supply_network#Water_distribution_network

http://statesborolandscape.com/statesboro-georgia-irrigation-systems

http://tlirr.com/products/center-pivot/#prettyPhoto/1/

http://www.env.gov.bc.ca/wat/wq/reference/glossary.html.

http://www.epa.gov/greeningepa/stormwater/

http://www.fao.org/docrep/s8684e/s8684e06.htm

http://www.thunderbirdirrigation.com

http://water-waysirrig.com/traveling-sprinklers

Irrigation Pipe on Wheels Move Across Fields, July 1950 *Popular Science*, bottom of page 114.

The Ride family's 'Nomad' brand tractor lawn sprinkler, National Museum of Australia.

GLOSSARY OF TERMS

Center pivot sprinklers: An automated irrigation system composed of a sprinkler lateral rotating around a pivot point and supported by a number of self-propelled towers. Water and power are supplied at the pivot point.

Gun systems: Uses a large sprinkler mounted on a wheeled cart or trailer, fed by a flexible rubber hose. The machine may be reeled in by the hose or moved periodically. Big guns operate at high operating pressures in order to throw water long distances.

LEPA (Low-Energy Precision Application): The use of drop tubes on a center pivot or lateral-move systems to apply water at low pressures very near or directly onto the soil surface. This increases efficiency and uniformity.

Nozzle: Final orifice through which water passes from the sprinkler to the atmosphere. Sprinkler flow rate is largely determined by the nozzle size (orifice diameter) and pressure.

Solid set: Refers to a stationary sprinkler system. Water-supply pipelines are generally fixed –usually below the soil surface – and sprinkler nozzles are elevated above the surface.

Sprinkler irrigation: Irrigation water is applied through a pressurized system. The pressure causes the water to flow out through the sprinkler nozzle and fly through the air and falls onto the soil surface. Commonly used agricultural sprinkler irrigation systems include center pivots, wheel lines, hand lines, solid set, and big guns.

Traveler: A sprinkler unit that propels itself along via the water pressure operating the sprinkler. Various types are manufactured which either operate via a pull wire or follow the layout of the hose supplying the water. Most units have an automatic shutoff at the end of the run.

Wheel line irrigation system: Also known as side-roll wheel move, or side move. Large-diameter wheels are mounted on a pipeline make it so that the pipeline can be rolled to the next position with less effort than hand-move. These are usually powered by a small gasoline engine.

14 Water reuse for irrigation

READING FOR COMPREHENSION

Water reuse is the process of reclaiming wastewater from a variety of sources and using it for useful purposes. Other terms that refer to water reuse are including water reclamation, water recycling, or wastewater reuse. The main reclaimed water applications are in the following sectors:

- **Agriculture:** Irrigation of agriculture fields, orchards, and greenhouses; animal husbandry; aquaculture.
- **Urban:** Irrigation of parks, private gardens, and sporting complexes; surface cleaning of streets, roads, roadsides, and other trafficked areas; fire protection systems; car washing; air conditioners; toilet flushing.
- **Industry:** Process water for power plants, refineries, mills, and factories; soil compaction; concrete mixing and other construction processes.
- **Recreation:** Fishing; boating; swimming; snowmaking; golf course irrigation; aesthetic.
- **Environment:** Creation of recharge water bodies including both surface (wetlands, artificial lakes, or streamflow) and sub-surface recourses (aquifers); dust control.
- **Potable uses:** Use of reclaimed water as drinking water after high levels of treatment (direct) or passing through the natural environment (indirect) (Figure 14.1).

WATER REUSE FOR AGRICULTURAL IRRIGATION

Benefits of Water Reuse in Agriculture

The water reuse in agriculture can provide significant advantages, directly or indirectly. Water reuse advantages can be environmental, economic, social, political, and security. Some of these advantages are listed as below:

- **Control the water scarcity:** Currently one-third of the world's population living in regions under water stress. Reclaimed water can make a significant contribution to meeting the world's water demands, decreasing overexploitation of surface and groundwater, and consequently reducing the human impact on the world's water environment.
- **Reduce the water conflicts:** Freshwater resources are vital, yet scarce, and unevenly distributed, therefore, a wide range of water conflicts appeared throughout history and disputes over water are predicted to be the source

DOI: 10.1201/9781003293507-14

FIGURE 14.1   Irrigation of wheat fields, Ahvaz, Iran (By Mohammad Albaji).

of future wars. Using water reclaimed, as an alternative for freshwater, can reduce the water conflicts.

- **Increase the food security:** In 2020, the FAO reported that the number of undernourished people was close to 690 million, or 8.9% of the world's population. One of the most important challenges to achieving food security is the global water crisis. Water reuse contributes to food security around the world by increasing agricultural production.
- **Energy saving:** Water and energy are intricately linked, and water reuse can reduce energy consumption through various ways. For example, as the groundwater level rises, the energy required for pumping decreases as well.
- **Reduce the impacts of wastewater disposal:** Wastewater is commonly discharged into surface water sources due to lack of or insufficient wastewater treatment facilities. The release of raw or improperly treated wastewater on water bodies has destructive effects on all components of the environment, including human, animal, and aquatic health. Using reclaimed water can reduce the wastewater discharge and nutrient loads into the receiving environment.
- **Reduce the application of fertilizers:** Irrigation with reclaimed water can decrease the need for fertilization due to the existing nutrients (such as nitrogen, potassium, and phosphorous) in reclaimed water, even after

FIGURE 14.2 Water Scarcity, Buffaloes in Hur Al-Azim Wetland at summer 2021, Hovizeh, Iran (By Mohammad Albaji).

the water treatment process. Many studies and farmers have confirmed that fact.

• **Improve the local economy:** Water reuse helps local people economy through creating employment in agriculture fields, animal husbandry, aquaculture, and by-products industry (Figure 14.2).

HEALTH RISKS OF WATER REUSE AGRICULTURE

The safety of using wastewater for agricultural irrigation is a global concern because in addition to its potential benefits, the use of wastewater can cause serious risks. The water reuse can threaten human health and environment if the necessary measures, such as properly treatment, are not applied before using the wastewater. There are two primary types of pathogens that can be found in wastewater used for agricultural irrigation; biological (Bacteria, helminths, protozoans, virus, Schistosoma) or chemical (Antibiotics, cyanobacterial toxins, heavy metals, phthalates and phenols, halogenated hydrocarbons, pesticides, and their residues). The types of pathogens and their concentrations in wastewater vary based on the sanitary and socioeconomic conditions of the region. In low-income communities, microbiological risks are the main concern while in higher-income communities, where microbiological contaminants are well controlled, chemical risks and emerging pollutants are more public concerning. Some water-borne diseases related to wastewater are presented in Table 14.1.

TABLE 14.1

Some of the diseases associated to wastewater

Disease	Cause
Typhoid fever	Salmonella typhi
Paratyphoid fever	Salmonella paratyphi
Gastroenteritis	Salmonella typhimurium
Cholera	Vibrio cholerae
Bacillary dysentery	Shigella dysenteriae
Amebiasis	Entamoeba histolytica
Giardiasis	Giardia duodenalis
Cryptosporidiosis	Cryptosporidium
Cyclosporiasis	Cyclospora cayetanensis
Infectious hepatitis	Hepatitis A
Gastroenteritis	Enterovirus, parvovirus, rotavirus
Infantile paralysis	Poliovirus
Leptospirosis	Leptospira icterohaemorrhagiae
Ear infections	Pseudomonas aeruginosa
Scabies	Sarcoptes scabiei
Trachoma	Chlamydia trachomatis
Schistosomiasis	Schistosoma
Malaria	Plasmodium
Yellow fever	Flavivirus
Dengue	Flavivirus

AGRICULTURAL WATER REUSE GUIDELINES AND REGULATIONS

Considering the possible risks related to use wastewater in agriculture, the international community recognizes the safety of water reuse in agriculture as a serious water resources issue. To address this issue, many national and international organizations (governmental or non-governmental) provided safety standards as regulations and guidelines. Although in some countries the regulations or guidelines are implemented, many other countries, especially the developing countries, are using wastewater in an unregulated manner. Failure to follow the safety standards of water reuse can cause health risks that may result in significant secondary impacts. Although the guidelines come from different perspectives and have different audiences, they are conceptually similar. Some of the most common and important standards of water reuse in agriculture are described below.

WHO GUIDELINES

In 1989, the World Health Organization (WHO) provided guidance on the microbiological quality for wastewater use in agriculture, considering the various health aspects of wastewater reuse. This recommendation is classified into three general categories based on the different reuse conditions and exposed groups:

A. Irrigation of crops likely to be eaten uncooked, sports field, public parks (Exposed group: Workers, consumers, public).
B. Irrigation of cereal crops, industrial crops, fodder crops pasture, and trees (Exposed group: Workers).
C. Localized irrigation of crops in category B (Exposed group: None).

For each of the above three categories, the maximum acceptable amount of intestinal nematodes and faucal coliforms, and the wastewater treatment expected to achieve the required microbiological quality are determined.

The WHO guidelines are described as reasonable minimum requirements of proper application of wastewater in agriculture but are not mandatory limits and how and to what extent they are implemented is left to the national governments. The WHO recommends that in special conditions, local factors in terms of epidemiology, cultural-social and environmental should be considered and the necessary adjustment should be made in the recommended table. For instance, WHO guidelines may be more sensitive to conditions in developing countries, and it would be expected that developed countries implement more stringent guidelines.

EPA GUIDELINES

In 1992, the United States Environmental Protection Agency (EPA) confirmed the toxic effects on crops irrigated with wastewater containing certain trace elements. In 2004, the EPA expanded their researches scope and produced comprehensive guidelines for water reuse. The EPA guidelines contained types of reuse applications, technical issues in planning water reuse systems, water reuse regulations and guidelines in the U.S, legal and institutional issues, funding water reuse systems, public involvement programs, and water reuse outside the U.S. The EPA's recommended limits for constituents in reclaimed water for irrigation take into account a wide range of elements (e.g. Aluminum, Iron, Manganese, Nickel, Zinc) and indexes (e.g. pH, TDS, free Chlorine residual). These limitations are presented separately for both short-term and long-term use. There are no federal requirements specific to implement EPA's guidelines for water reuse; states have that authority.

EUROPEAN UNION REGULATIONS

The water resources of the European Union are increasingly under pressure, leading to water shortages and declining water quality. One way to address this situation is the wider reuse of wastewater, which requires the creation of an instrument to regulate standards at Union level. For this aim, the European Union has set regulations on minimum water reuse requirements for irrigation purposes since 2020. The EU standard for the use of wastewater in irrigation divides plants into four categories and determines for each category the minimum reclaimed water quality using E. coli, BOD, TSS, and Turbidity indices. In addition, the EU standard provides regulations of the minimum frequencies for routine monitoring and validation monitoring of reclaimed water for agricultural irrigation. This regulation shall enter into force from June 26, 2023, and directly applicable in all Member States. Member States should take all necessary actions to make sure of regulation implementation, and lay down the rules on penalties to infringements of this regulation.

EXERCISES

A. **Read each statement and decide whether it is true or false. Write "T" for true and "F" for false statements.**

TF1. Reclaimed wastewater cannot be used for aquaculture.

TF2. Using wastewater can cause energy saving.

TF3. There is no nutrient in the water after the treatment process.

TF4. Already the international community recognizes the safety of water reuse in agriculture as a serious water resources issue.

TF5. Member States of the European Union should lay down the rules on penalties to infringements of this regulation.

B. **Circle a, b, c, or d which best completes the following items.**

1. Reclaimed water can be used in .... for soil compaction and process water for power plants.
   a. urban
   b. industry
   c. environment
   d. recreation

2. Water reuse can control the .... as one of the most important threats to food security in the world.
   a. lack of energy
   b. water scarcity
   c. agricultural land restrictions
   d. the application of fertilizers

3. .... are chemical pathogen that may exist in wastewater used for agricultural irrigation.
   a. Helminthes
   b. Phthalates
   c. Protozoans
   d. Schistosoma

4. .... is not a disease associated with the use of wastewater.
   a. Cholera
   b. Leptospirosis
   c. Flu
   d. Giardiasis

5. in the standard developed by .... for the use of wastewater, plants are divided into four categories, for each of which the minimum reclaimed water quality are determinate.
   a. FAO
   b. WHO
   c. EPA
   d. European Union

C. **Match the sentence halves in Column I with their appropriate halves in Column II. Insert the letters a, b, c ... in the parentheses provided. There are more sentence halves in Column II than required.**

Column I	Column II
1. Turbidity	( ) **a.** the science of breeding animals for their products.
2. Orchard	( ) **b.** reclaiming wastewater from different sources and using it for useful intentions.
3. Aquaculture	( ) **c.** any organism that can produce disease.
4. Water reuse	( ) **d.** the cloudiness or haziness of a fluid caused by large numbers of individual particles.
5. Animal husbandry	( ) **e.** creation or recharge water bodies.
	( ) **f.** the farming of aquatic animals and plants for food.
	( ) **g.** an intentional plantation of trees or shrubs for food production.
	( ) **h.** increasing agricultural production for food security.

D. **Cross out the word or words that make each statement false, and write the word or words that make each statement true in the blank.**

1. In high-income countries, risks from microbiological pollution is the main concern about wastewater reuse while in low-income countries, chemical pollution is more important.

   ....

2. The pathogens of Typhoid fever and Malaria are Flavivirus and Salmonella paratyphi, respectively.

   ....

3. Reclaimed water can be used as potable water after primary treatment.

   ....

4. The WHO guidelines for water reuse shall enter into force in all Member States while the European Union regulations are not mandatory limits.

   ....

E. **Give answers to the following questions.**
1. What are the main reclaimed water uses?
2. What are the benefits of water reuse in agriculture?
3. Why using reclaimed water for irrigation reduce the application of fertilizers?
4. What are the most famous agricultural water reuse guidelines and regulations?

F. **For each word on the left, there are three meanings provided. Put a check mark (√) next to the choice which has the closest meaning to the word given.**

1. **Potable water**	wastewater	drinking water	treated water
2. **Aquaculture**	aquatic ecosystem	aquifers	aquafarming
3. **Water reclamation**	water conflict	water crisis	wastewater reuse
4. **Water crisis**	water body	water scarcity	water management
5. **Pathogen**	water nutrient	water components	disease cause

G. **Fill in the blanks with the appropriate words from the following list.**

faecal coliforms   local economy   Turbidity   safety standards
exposed group   water conflicts   potable water   Cholera

The wastewater can be recycled for reuse in various sectors including agriculture, urban, industry, recreation, and environment. The wastewater can even be used as .... after high levels of treatment. Water reuse for agricultural irrigation causes considerable benefits in economic, social, political, and security fields. Water reclaiming contributes to control the water scarcity, reduce the ...., increase the food security, energy saving, reduce the impacts of wastewater disposal, reduce the application of fertilizers, and improve the .... Despite the mentioned benefits, the use of wastewater can threaten the human health and the environment. The biological and chemical pathogens that can be found in untreated or inadequately treated wastewater can cause diseases such as Typhoid fever, Paratyphoid fever, ..., Giardiasis, and Malaria. To address the associated risks with wastewater reuse, many national and international organizations developed .... as regulations and guidelines. The WHO guidance for wastewater use in agriculture determines the maximum acceptable amount of intestinal nematodes and .... for three general categories based on the different reuse conditions and .... The EU standard is classified into four categories based on the plants and determines for each category the minimum reclaimed water quality using *E. coli*, BOD, TSS, and .... indices.

H. **Read this passage and then circle a, b, c, or d which best completes the following items.**

Smallholder farmers in developing countries should take actions against potential negative impacts on the environment caused by irrigation with untreated or inadequately treated wastewater. They can use management options to address the challenges and risks of exposure to elevated levels of metals, metalloids, salts, specific ionic species and added nutrients. These interventions include crop, soil, and water-based options.

There are the following solutions to risk management of metals and metalloids:

• Identify geographical areas with elevated risks from specific metal sources.
• Perform quality-assured testing of soil and plant samples to verify the level of the risk from specific metal(s).
• Identify alternative crop varieties of the same desired crop that take up the least metal or convert the toxin to less toxic forms when grown in high-risk areas.
• Develop irrigation, fertilization, and residue management strategies that help to minimize metal uptake by plants.
• Recommend crop restrictions, i.e. using other crops that have lower risks of contamination with metals and metalloids and/or pose a lesser risk to human health due to levels of dietary intake.
• Zone the affected area(s) for non-agricultural land use or land rehabilitation.

Nevertheless, in assessing environmental risk management in developing countries, there is often insufficient analytical capacity to analyze specific heavy metals and, in particular, organic contaminants.

Smallholder farmers in developing countries ....
a. have no adequate capacity to safe use of wastewater.
b. can contribute to the safe and effective use of wastewater through their interventions.
c. are using untreated wastewater for irrigation without negative impacts on the environment.
d. should choose plants that have the maximum metal uptake to clear agricultural land.

I. **Translate the following passage into your mother language. Write your translation in the space provided.**

Recent water crisis and droughts have reminded us the importance of the protection of the environment and public health by reliable and efficient water systems. One of the most important issues in water-related projects is financing. The U.S Environmental Protection Agency's (EPA) Clean Water State Revolving Fund (CWSRF) program is the largest public source of water quality projects financing in the USA. CWSRF programs in each state operate like banks, federal and state assets are used to pay low-interest loans for important water quality projects. Then the funds are reimbursed to the CWSRF and recycled to fund other water quality and public health projects.

States use priority-setting systems as an effective tool to encourage water reuse and conservation. CWSRF program evaluates and ranks projects using the priority system. The most important criteria for ranking are public health and water quality, but it can also consider other priorities, such as water reuse and conservation. Based on this prioritization, states may encourage some projects through offering priority points, and funding incentives, including reduced interest rates and waiver of fees.

The CWSRF considers a wide range of water reuse and conservation projects to be eligible for funding, including:
• Equipment to reuse reclaimed water (public and private)
• Direct potable reuse (public and private)
• Installation of water-efficient appliances and irrigation equipment
• Installation or replacement of water meters
• Plumbing fixtures and retrofit replacements

CWSRF programs are not limited to mentioning activities and they can also provide assistance to many types of planning activities.

J. **Copy the technical terms and expressions used in this lesson. Then find your mother language equivalents of those terms and expressions and write them in the spaces provided.**

Technical term	Mother language equivalent
....	....
....	....
....	....
....	....
....	....
....	....
....	....
....	....
....	....
....	....

BIBLIOGRAPHY

Anderson, J. (2003). The environmental benefits of water recycling and reuse. *Water Science and Technology: Water Supply* 3(4): 1–10

Edokpayi, J. N., Odiyo, J. O., & Durowoju, O. S. (2017). Impact of wastewater on surface water quality in developing countries: a case study of South Africa. Water quality, 401–416.

http://conservewaterforfood.org/water-reuse-in-agriculture

https://en.wikipedia.org/wiki/Animal_husbandry

https://en.wikipedia.org/wiki/Aquaculture

https://en.wikipedia.org/wiki/Fecal_coliform

https://en.wikipedia.org/wiki/Food_security

https://en.wikipedia.org/wiki/Orchard

https://en.wikipedia.org/wiki/Pathogen

https://en.wikipedia.org/wiki/Reclaimed_water

https://en.wikipedia.org/wiki/Water_conflict

https://images.app.goo.gl/9Bwn3rkzcPmiPtc28

https://images.app.goo.gl/BTqxukjFZNm3ET8E8

https://images.app.goo.gl/V6DqzCLtN8wTDaHX8

https://www.epa.gov/cwsrf

https://www.epa.gov/waterreuse/basic-information-about-water-reuse

https://www.lacsd.org/waterreuse/benefits.asp

https://www.watertechonline.com/wastewater/article/15550537/agricultural-water-reuse-and-guidelines

Jaramillo, M. F. & Restrepo, I. (2017). Wastewater reuse in agriculture: A review about its limitations and benefits. *Sustainability* 9(10): 1734.

Kretschmer, N., Ribbe, L., & Gaese, H. (2002). Wastewater reuse for agriculture. *Technology Resource Management & Development-Scientific Contributions for Sustainable Development* 2: 37–64.

Mateo-Sagasta, J., Medlicott, K., Qadir, M., Raschid-Sally, L., & Drechsel, P. (2013). Proceedings of the UN-water project on the safe use of wastewater in agriculture.

Regulation (EU) 2020/741 of the European Parliament and of the Council of 25 May 2020 on minimum requirements for water reuse. *Official Journal of the European Union.* https://eur-lex.europa.eu/legal-content/EN/TXT/?uri=CELEX%3A32020R0741.

The State of Food Security and Nutrition in the World 2020, *In brief.* Rome: FAO, IFAD, UNICEF, WFP and WHO. 2020. p. 12. 10.4060/ca9699en. ISBN 978-92-5-132910-8.

Ungureanu, N., Vlăduț, V., & Voicu, G. (2020). Water Scarcity and Wastewater Reuse in Crop Irrigation. *Sustainability* 12(21): 9055.

United States. Environmental Protection Agency. Office of Wastewater Management. Municipal Support Division, National Risk Management Research Laboratory (US). Technology Transfer, & Support Division. (2004). *Guidelines for water reuse.* US Environmental Protection Agency.

Vörösmarty, C. J., McIntyre, P. B., Gessner, M. O., Dudgeon, D., Prusevich, A., Green, P.,... & Davies, P. M. (2010). Global threats to human water security and river biodiversity. *Nature* 467(7315): 555–561.

World Health Organization. (1989). *Health guidelines for the use of wastewater in agriculture and aquaculture: report of a WHO scientific group [meeting held in Geneva from 18 to 23 November 1987].* World Health Organization.

World Health Organization. (2006). *WHO guidelines for the safe use of wasterwater excreta and greywater (Vol. 1).* World Health Organization.

GLOSSARY OF TERMS

Animal husbandry: The branch of agriculture concerned with animals that are raised for meat, fiber, milk, eggs, or other products.

Aquaculture: The farming of fish, crustaceans, mollusks, aquatic plants, algae, and other organisms.

Fecal coliform: A facultative anaerobic, rod-shaped, gram-negative, non-sporulating bacterium.

Orchard: An intentional plantation of trees or shrubs, which are generally grown for commercial food production.

Pathogen: Any organism that can produce disease.

Potable Water: Water that is suitable for human consumption (i.e., water that can be used for drinking or cooking).

Turbidity: The cloudiness or haziness of a fluid caused by large numbers of individual particles that are generally invisible to the naked eye.

Water conflicts: A conflict between countries, states, or groups over the rights to access water resources.

15 Irrigation system and structures

READING FOR COMPREHENSION

The irrigation system consists of a (main) intake structure or (main) pumping station, a conveyance system, a distribution system, a field application system, and a drainage system (Figure 15.1).

The (main) intake structure, or (main) pumping station, directs water from the source of supply, such as a reservoir or a river, into the irrigation system.

The conveyance system assures the transport of water from the main intake structure or main pumping station up to the field ditches.

The distribution system assures the transport of water through field ditches to the irrigated fields.

The field application system assures the transport of water within the fields.

The drainage system removes the excess water (caused by rainfall and/or irrigation) from the fields.

MAIN INTAKE STRUCTURE AND PUMPING STATION

Main intake structure

The intake structure is built at the entry to the irrigation system. Its purpose is to direct water from the original source of supply (lake, river, reservoir, etc.) into the irrigation system (Figure 15.2).

PUMPING STATION

In some cases, the irrigation water source lies below the level of the irrigated fields. Then a pump must be used to supply water to the irrigation system (Figure 15.3).

There are several types of pumps, but the most commonly used in irrigation is the centrifugal pump.

The centrifugal pump consists of a case in which an element, called an impeller, rotates driven by a motor. Water enters the case at the center, through the suction pipe. The water is immediately caught by the rapidly rotating impeller and expelled through the discharge pipe (Figures 15.4 and 15.5).

The centrifugal pump will only operate when the case is completely filled with water.

CONVEYANCE AND DISTRIBUTION SYSTEM

The conveyance and distribution systems consist of canals transporting the water through the whole irrigation system. Canal structures are required for the control and measurement of the water flow.

DOI: 10.1201/9781003293507-15

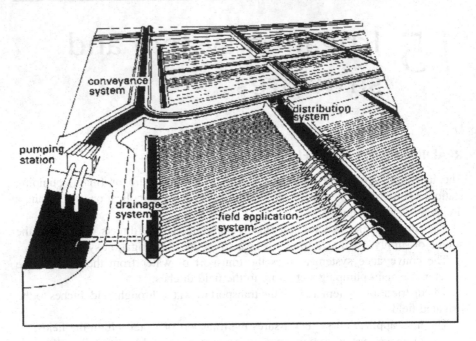

FIGURE 15.1   An irrigation system; (http://www.fao.org/docrep/r4082e/r4082e06.htm).

OPEN CANALS

An open canal, channel, or ditch, is an open waterway whose purpose is to carry water from one place to another. Channels and canals refer to main waterways supplying water to one or more farms. Field ditches have smaller dimensions and convey water from the farm entrance to the irrigated fields.

CANAL CHARACTERISTICS

According to the shape of their cross-section, canals are called rectangular (a), triangular (b), trapezoidal (c), circular (d), parabolic (e), and irregular or natural (f) (Figure 15.6).

The most commonly used canal cross-section in irrigation and drainage is the trapezoidal cross-section. For the purposes of this publication, only this type of canal will be considered.

The typical cross-section of a trapezoidal canal is shown in Figure 15.7.

The freeboard of the canal is the height of the bank above the highest water level anticipated. It is required to guard against overtopping by waves or unexpected rises in the water level.

The side slope of the canal is expressed as ratio, namely the vertical distance or height to the horizontal distance or width. For example, if the side slope of the canal has a ratio of 1:2 (one to two), this means that the horizontal distance (w) is two times the vertical distance (h) (Figure 15.9).

FIGURE 15.2 Intake structure, Karkeh River, Iran (Photo by: Abdullah Asakereh).

The bottom slope of the canal does not appear on the drawing of the cross-section but on the longitudinal section. It is commonly expressed in percent or per mil (Figure 15.10).

An example of the calculation of the bottom slope of a canal is given below (see also Figure 15.8):

$$\text{the bottom slope}(\%) = \frac{\text{height difference (metres)}}{\text{horizontal distance (metres)}} \times 100 = \frac{1\text{m}}{100\text{ m}} \times 100$$
$$= 1\%$$

or

$$\text{the bottom slope}(\%) = \frac{\text{height difference (metres)}}{\text{horizontal distance (metres)}} \times 1000 = \frac{1\text{m}}{100\text{ m}} \times 1000$$
$$= 10\%$$

FIGURE 15.3   Kowsar Pumping Station, Karkeh River, Iran (Photo by: Abdullah Asakereh).

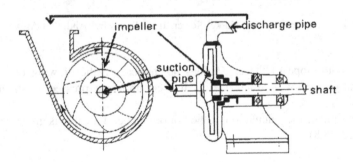

FIGURE 15.4   Diagram of a centrifugal pump; (http://www.fao.org/docrep/r4082e/r4082e06.htm).

Earthen Canals

Earthen canals are simply dug in the ground and the bank is made up from the removed earth.

The disadvantages of earthen canals are the risk of the side slopes collapsing and the water loss due to seepage. They also require continuous maintenance in order to control weed growth and to repair the damage done by livestock and rodents (Figure 15.11).

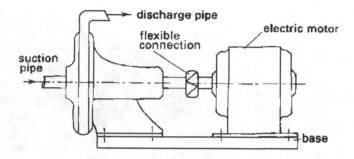

FIGURE 15.5 Centrifugal pump and motor; (http://www.fao.org/docrep/r4082e/r4082e06.htm).

FIGURE 15.6 A rectangular canal, Shahid Chamran University of Ahvaz, Iran (By Mohammad Albaji).

LINED CANALS

Earthen canals can be lined with impermeable materials to prevent excessive seepage and growth of weeds.

Lining canals is also an effective way to control canal bottom and bank erosion. The materials mostly used for canal lining are concrete (in precast slabs or cast in place), brick or rock masonry and asphaltic concrete (a mixture of sand, gravel, and asphalt).

FIGURE 15.7 A trapezoidal canal, Shahid Chamran University of Ahvaz, Iran (By Mohammad Albaji).

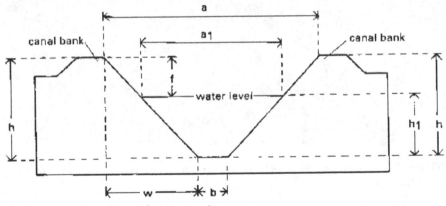

a   = top width of the canal
a_1 = top width of the water level
h   = height of the canal
h_1 = heigth or depth of the water in the canal
b   = bottom width of the canal
h:w = side slope of the canal
f   = free board (= $h-h_1$)

FIGURE 15.8 A trapezoidal canal cross-section; (http://www.fao.org/docrep/r4082e/r4082e06.htm).

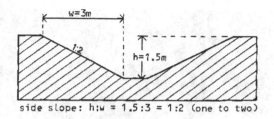

FIGURE 15.9 A side slope of 1:2 (one to two); (http://www.fao.org/docrep/r4082e/r4082e06.htm).

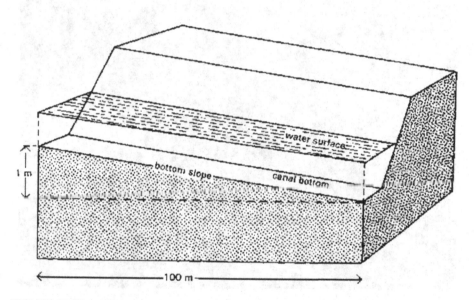

FIGURE 15.10 A bottom slope of a canal; (http://www.fao.org/docrep/r4082e/r4082e06.htm).

The construction cost is much higher than for earthen canals. Maintenance is reduced for lined canals but skilled labor is required.

CANAL STRUCTURES

The flow of irrigation water in the canals must always be under control. For this purpose, canal structures are required. They help regulate the flow and deliver the correct amount of water to the different branches of the system and onward to the irrigated fields.

There are four main types of structures: erosion control structures, distribution control structures, crossing structures, and water measurement structures.

FIGURE 15.11   Earthen canal, Shahid Chamran University of Ahvaz, Iran (By Mohammad Albaji).

EROSION CONTROL STRUCTURES

Canal erosion

Canal bottom slope and water velocity are closely related, as the following example will show.

A cardboard sheet is lifted on one side 2 cm from the ground. A small ball is placed at the edge of the lifted side of the sheet. It starts rolling downward, following the slope direction. The sheet edge is now lifted 5 cm from the ground, creating a steeper slope. The same ball placed on the top edge of the sheet rolls downward, but this time much faster. The steeper the slope, the higher the velocity of the ball (Figure 15.13).

FIGURE 15.12   Lined canal, Hamedan, Iran (By Mehdi Jovzi).

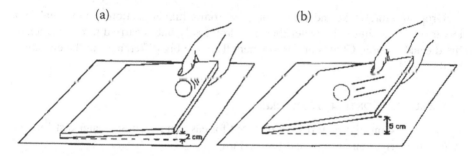

FIGURE 15.13   The relationship between slope and velocity; (http://www.fao.org/docrep/r4082e/r4082e06.htm).

Water poured on the top edge of the sheet reacts exactly the same as the ball. It flows downward and the steeper the slope, the higher the velocity of the flow.

Water flowing in steep canals can reach very high velocities. Soil particles along the bottom and banks of an earthen canal are then lifted, carried away by the water flow, and deposited downstream where they may block the canal and silt up structures. The canal is said to be under erosion; the banks might eventually collapse.

DROP STRUCTURES AND CHUTES

Drop structures or chutes are required to reduce the bottom slope of canals lying on steeply sloping land in order to avoid high velocity of the flow and risk of erosion. These structures permit the canal to be constructed as a series of relatively flat sections, each at a different elevation (Figure 15.14).

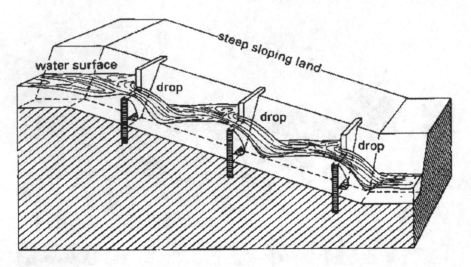

FIGURE 15.14 Longitudinal section of a series of drop structures; (http://www.fao.org/docrep/r4082e/r4082e06.htm).

Drop structures take the water abruptly from a higher section of the canal to a lower one. In a chute, the water does not drop freely but is carried through a steep, lined canal section. Chutes are used where there are big differences in the elevation of the canal.

Distribution Control Structures

Distribution control structures are required for easy and accurate water distribution within the irrigation system and on the farm.

Division Boxes

Division boxes are used to divide or direct the flow of water between two or more canals or ditches. Water enters the box through an opening on one side and flows out through openings on the other sides. These openings are equipped with gates (Figure 15.15).

Turnouts are constructed on the bank of a canal. They divert part of the water from the canal to a smaller one.

Turnouts can be concrete structures pipe structures (Figure 15.12) (Figure 15.16).

Checks

To divert water from the field ditch to the field, it is often necessary to raise the water level in the ditch. Checks are structures placed across the ditch to block it temporarily and to raise the upstream water level. Checks can be permanent structures or portable (Figure 15.17).

(a)

(b)

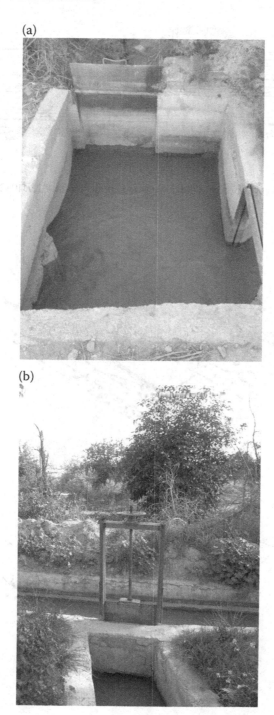

FIGURE 15.15   a. A division box with two gates, Shahid Chamran University of Ahvaz, Iran.
b. An Irrigation gate, Shahid Chamran University of Ahvaz, Iran (By Mohammad Albaji).

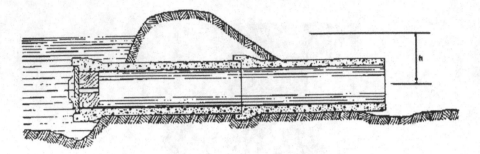

FIGURE 15.16   A pipe turnout; (http://www.fao.org/docrep/r4082e/r4082e06.htm).

(a)

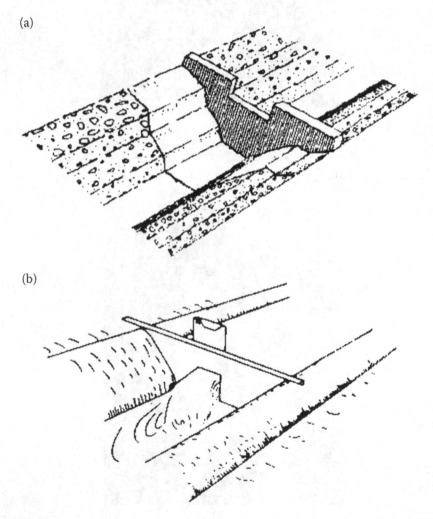

(b)

FIGURE 15.17   a. A permanent concrete check; (http://www.fao.org/docrep/r4082e/
r4082e06.htm). **b.** A portable metal check; (http://www.fao.org/docrep/r4082e/r4082e06.htm).

CROSSING STRUCTURES

It is often necessary to carry irrigation water across roads, hillsides, and natural depressions. Crossing structures, such as flumes, culverts, and inverted siphons, are then required.

FLUMES

Flumes are used to carry irrigation water across gullies, ravines, or other natural depressions. They are open canals made of wood (bamboo), metal, or concrete which often need to be supported by pillars.

CULVERTS

Culverts are used to carry the water across roads. The structure consists of masonry or concrete headwalls at the inlet and outlet connected by a buried pipeline (Figure 15.18).

INVERTED SIPHONS

When water has to be carried across a road that is at the same level as or below the canal bottom, an inverted siphon is used instead of a culvert. The structure consists of an inlet and outlet connected by a pipeline. Inverted siphons are also used to carry water across wide depressions (Figures 15.19 and 15.20).

FIGURE 15.18 Culverts, Hamedan, Iran (By Mehdi Jovzi).

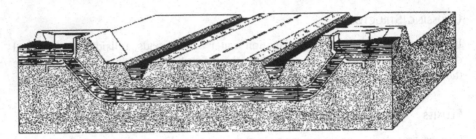

FIGURE 15.19   An inverted siphon; (http://www.fao.org/docrep/r4082e/r4082e06.htm).

FIGURE 15.20   Floating the Karun River inverted siphon, Abadan, Iran (By Faramarez Ghalambaz).

WATER MEASUREMENT STRUCTURES

The principal objective of measuring irrigation water is to permit efficient distribution and application. By measuring the flow of water, a farmer knows how much water is applied during each irrigation.

In irrigation schemes where water costs are charged to the farmer, water measurement provides a basis for estimating water charges.

The most commonly used water measuring structures are weirs and flumes. In these structures, the water depth is read on a scale that is part of the structure. Using this reading, the flow rate is then computed from standard formulas or obtained from standard tables prepared specially for the structure.

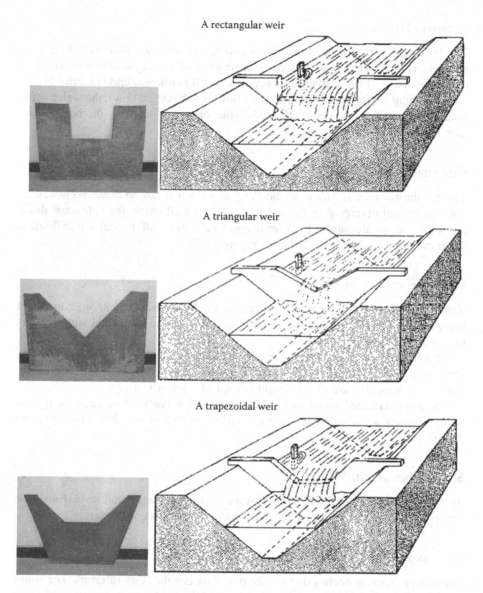

FIGURE 15.21 Some examples of weirs; (http://www.fao.org/docrep/r4082e/r4082e06.htm).

Weirs

In its simplest form, a weir consists of a wall of timber, metal, or concrete with an opening with fixed dimensions cut in its edge. The opening, called a notch, may be rectangular, trapezoidal, or triangular (Figure 15.21).

PARSHALL FLUMES

The Parshall flume consists of a metal or concrete channel structure with three main sections: (1) a converging section at the upstream end, leading to (2) a constricted or throat section, and (3) a diverging section at the downstream end (Figure 15.22).

Depending on the flow condition (free flow or submerged flow), the water depth readings are taken on one scale only (the upstream one) or on both scales simultaneously.

CUT-THROAT FLUME

The cut-throat flume is similar to the Parshall flume but has no throat section, only converging and diverging sections. Unlike the Parshall flume, the cut-throat flume has a flat bottom. Because it is easier to construct and install, the cut-throat flume is often preferred to the Parshall flume (Figure 15.23).

FIELD APPLICATION SYSTEMS

There are many methods of applying water to the field. The simplest one consists of bringing water from the source of supply, such as a well, to each plant with a bucket or a water can (Figure 15.24).

This is a very time-consuming method and it involves quite heavy work. However, it can be used successfully to irrigate small plots of land, such as vegetable gardens, that are in the neighborhood of a water source.

More sophisticated methods of water application are used in larger irrigation systems. There are three basic methods: surface irrigation, sprinkler irrigation, and drip irrigation.

SURFACE IRRIGATION

Surface irrigation is the application of water to the fields at ground level. Either the entire field is flooded or the water is directed into furrows or borders.

FURROW IRRIGATION

Furrows are narrow ditches dug on the field between the rows of crops. The water runs along them as it moves down the slope of the field.

The water flows from the field ditch into the furrows by opening up the bank or dyke of the ditch or by means of siphons or spiles. Siphons are small curved pipes that deliver water over the ditch bank (see Figure 15.25). Spiles are small pipes buried in the ditch bank.

BORDER IRRIGATION

In border irrigation, the field to be irrigated is divided into strips (also called borders or border strips) by parallel dykes or border ridges.

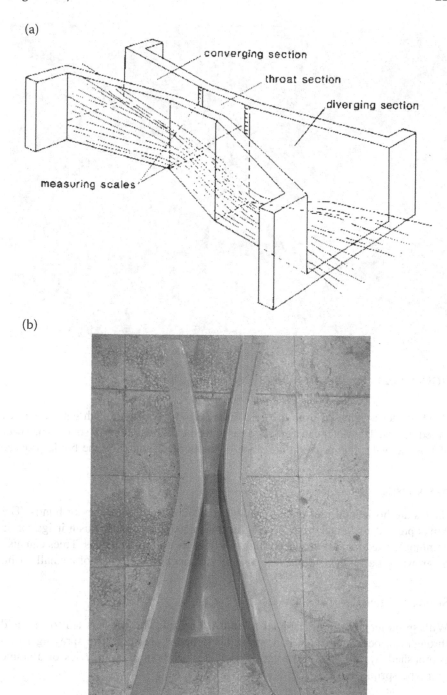

FIGURE 15.22 A Parshall flume; (http://www.fao.org/docrep/r4082e/r4082e06.htm).

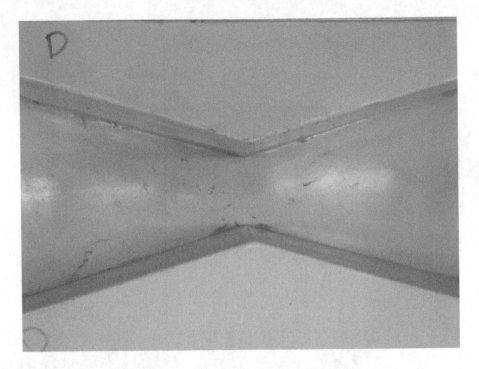

FIGURE 15.23   A cut-throat flume (By Mohammad Albaji).

The water is released from the field ditch onto the border through gate structures called outlets. The water can also be released by means of siphons or spiles. The sheet of flowing water moves down the slope of the border, guided by the border ridges.

BASIN IRRIGATION

Basins are horizontal, flat plots of land, surrounded by small dykes or bunds. The banks prevent the water from flowing to the surrounding fields. Basin irrigation is commonly used for rice grown on flatlands or in terraces on hillsides. Trees can also be grown in basins, where one tree usually is located in the center of a small basin.

SPRINKLER IRRIGATION

With sprinkler irrigation, artificial rainfall is created. The water is led to the field through a pipe system in which the water is under pressure. The spraying is accomplished by using several rotating sprinkler heads or spray nozzles or a single gun type sprinkler.

DRIP IRRIGATION

In drip irrigation, also called trickle irrigation, the water is led to the field through a pipe system. On the field, next to the row of plants or trees, a tube is installed. At

FIGURE 15.24 Manual irrigation with water canes. Shahid Chamran University of Ahvaz, Iran (By Mohammad Albaji).

(a)   (b)

(c)

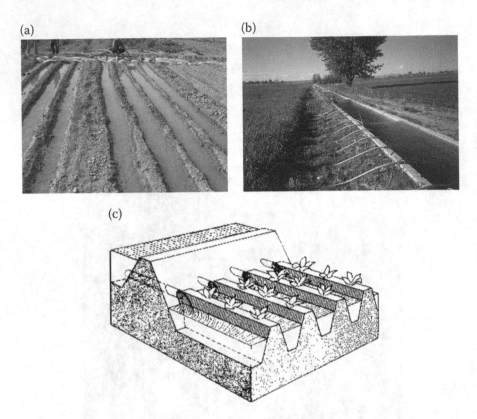

FIGURE 15.25 a. Water flows into the furrows through openings in the bank, Shahid Chamran University of Ahvaz, Iran. **b.** The use of siphons (https://en.wikipedia.org/wiki/Surface_irrigation#/media/File:SiphonTubes.JPG). **c.** The use of spiles; (http://www.fao.org/docrep/r4082e/r4082e06.htm).

regular intervals, near the plants or trees, a hole is made in the tube and equipped with an emitter. The water is supplied slowly, drop by drop, to the plants through these emitters.

DRAINAGE SYSTEM

A drainage system is necessary to remove excess water from the irrigated land. This excess water may be e.g. wastewater from irrigation or surface runoff from rainfall. It may also include leakage or seepage of water from the distribution system.

Excess surface water is removed through shallow open drains. Excess groundwater is removed through deep open drains or underground pipes.

EXERCISES

A. **Read each statement and decide whether it is true or false. Write "T" for true and "F" for false statements.**

TF1. The main pumping station is built at the entry to the irrigation system.

TF2. If the irrigation water source lies over the level of the irrigated fields, then a pump must be used to supply water to the irrigation system.

TF3. The most commonly used canal in irrigation is the trapezoidal cross-section.

TF4. The side slope of the canal is expressed as ratio, namely the width to the height.

TF5. The flow of irrigation water in the canals must always be under control.

B. **Circle a, b, c, or d which best completes the following items.**

1. The .... assures the transport of water through field ditches to the irrigated fields.
   a. intake structure
   b. distribution system
   c. conveyance system
   d. pumping station

2. .... are used to carry the water across roads.
   a. Inverted siphons
   b. Flumes
   c. Culverts
   d. Parshall flume

3. .... can be lined with impermeable materials to prevent excessive seepage and growth of weeds.
   a. Flumes
   b. Division box
   c. Earthen canals
   d. Lined canal

4. .... are used to carry irrigation water across gullies, ravines, or other natural depressions.
   a. Flumes
   b. Culverts
   c. Inverted siphons
   d. Weirs

5. The .... systems consist of canals transporting the water through the whole irrigation system.
   a. conveyance and distribution
   b. conveyance and drainage
   c. conveyance and field application
   d. field application and distribution

C. **Match the sentence halves in Column I with their appropriate halves in Column II. Insert the letters a, b, c ... in the parentheses provided. There are more sentence halves in Column II than required.**

Column I	Column II
1. The field application system assures the	( ) **a.** transport of water through field ditches to the irrigated fields.
2. The conveyance system assures the	( ) **b.** removes the excess water from the fields.
3. The intake structure	( ) **c.** transport of water from the main pumping up to station the field ditches.
4. An open ditch,	( ) **d.** transport of water within the fields.
5. The freeboard of the canal	( ) **e.** directs water from the source of supply, into the irrigation system.
	( ) **f.** is the height of the bank above the highest water level anticipated.
	( ) **g.** consist of canals transporting the water through the whole irrigation system.
	( ) **h.** is an open waterway whose purpose is to carry water from one place to another.

D. **Cross out the word or words that make each statement false, and write the word or words that make each statement true in the blank.**

1. When water has to be carried across a road that is at the same level as or below the canal bottom, a culvert is used instead of an inverted siphon.

   ....

2. It is often necessary to carry irrigation water across roads, hillsides, and natural depressions. Distribution control structures, such as flumes, culverts, and inverted siphons, are then required.

   ....

3. Checks are constructed in the bank of a canal. They divert part of the water from the canal to a smaller one.

   ....

4. In its simplest form, a flume consists of a wall of timber, metal, or concrete with an opening with fixed dimensions cut in its edge.

   ....

E. **Give answers to the following questions.**

1. What is the most commonly used pump in irrigation?
2. What are the types of canals? (According to the shape of their cross-section)
3. What are the main types of irrigation structures?

F. **For each word on the left, there are three meanings provided. Put a check mark (√) next to the choice which has the closest meaning to the word given.**

---

1. **Drop**   chute   turnout   weir

2. **Cut-throat flume**   Parshall flume   weir   inverted siphons

3. **Conveyance system**   transport system   pumping system   distribution
                                                                     system

4. **Field application**   distribution   surface irrigation over irrigation
   systems   system

5. **Intake structure**   conveyance   distribution   pumping station
                             system   system

G. **Fill in the blanks with the appropriate words following from the following list.**

   committees unlined canal maintenance lining material
            siltation particles lining susceptible

A surface ...., such as concrete, brick, or plastic, on the canal prevents the growth of plants and discourages hole-making by rats or termites, and so the .... of a lined canal can be easier and quicker than that of an .... Moreover, the higher velocity that can safely be allowed in the lined canal prevents the small .... of soil carried in the water from settling out, accumulating, and causing .... The bed and sides of lined canals are more stable than those of unlined canals and are thus less .... to erosion.

The costs of lining can be very high, depending on the local cost of ....and of labor, as well as on the length of canal to be lined. Prices of lining material vary from place to place. Irrigation .... and farmers who are considering lining the canals in their irrigation scheme should gather information on prices of material and of the labor required.

H. **Read this passage and then circle a, b, c, or d which best completes the following items.**

Before the decision is made to line a canal, the costs and benefits of lining have to be compared. By lining the canal, the velocity of the flow can increase because of the smooth canal surface. For example, with the same canal bed slope and with the same canal size, the flow velocity in a lined canal can be 1.5 to 2 times that in an unlined canal, which means that the canal cross-section in the lined canal can be smaller to deliver the same discharge.

Possible benefits of lining a canal include: water conservation, no seepage of water into adjacent land or roads; reduced canal dimensions; and reduced maintenance.

Lining is said to be useful due to ....

a. the velocity of the flow can increase.

b. reduced canal dimensions.

c. reduced canal maintenance.

d. a, b, c are correct

I. **Translate the following passage into your mother language. Write your translation in the space provided.**

The choice of lining material depends primarily on: local costs; availability of materials; and availability of local skills (local craftsmen).

If cement, gravel, and sand are relatively cheap and locally available, concrete lining is generally a good choice. Although the initial investment in concrete lining is generally high, if it is properly constructed and maintained it could last for many years, which thus offsets the high initial cost. If a local fired brick industry produces cheap bricks or if construction stone or precast concrete slabs are locally available, brick or stone masonry or a concrete slab can be considered. Large amounts of cement are required for mortar and plastering. The construction of this type of lining requires more labor than other methods, thus its use tends to be limited to where labor is abundant and the material cost is relatively low.

If a sufficient volume of heavy clay is available near the irrigation scheme, a clay lining could be considered. Lining canals with clay is rather labor intensive, and so the costs of labor should be taken into account when comparing costs and benefits. The use of clay can reduce seepage losses and improve the smoothness of the canal surface, but does not stop weed growth and possible erosion. If coarse aggregates are not available and cement is relatively cheap, soil (sand) cement lining could be considered (Figure 15.26).

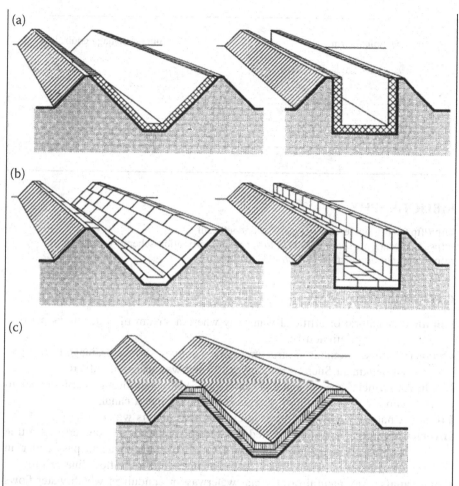

FIGURE 15.26 Different types of lining; (http://ftp.fao.org/docrep/fao/010/ai585e/ai585e04.pdf).

J. **Copy the technical terms and expressions used in this lesson. Then find your mother language equivalents of those terms and expressions and write them in the spaces provided.**

Technical term	Mother language equivalent
....	....
....	....
....	....
....	....
....	....

(*Continued*)

Technical term	Mother language equivalent
....	....
....	....
....	....
....	....
....	....

BIBLIOGRAPHY

http://ftp.fao.org/docrep/fao/010/ai585e/ai585e04.pdf
https://geolocation.ws/v/P/56490026/lower-waitaki-irrigation-scheme-waitaki/en
http://www.fao.org/docrep/r4082e/r4082e06.htm

GLOSSARY OF TERMS

Channel: A natural or artificial waterway where a stream of water flows periodically or continuously.

Culvert: A closed conduit or structure used to convey surface drainage through an embankment Such as a roadway. In highway usage, a culvert.

Ditch: An artificial open channel or waterway constructed through earth or rock to convey water. A ditch is generally smaller than a canal.

Flume: A natural or artificially made channel that diverts water.

Inverted Siphon: (Sometimes called a depressed sewer or sag culvert, not a true siphon). A structure, generally a length of pipe, made to pass under an obstruction in such a manner that a concavity in the flow line results.

Open canals: Any natural or artificial waterway or conduit in which water flows with a free surface.

Outlet: Downstream opening or discharge end of a pipe, culvert, ditch, or canal.

Parshall flume: Parshall flume is a fixed hydraulic structure developed to measure surface waters and irrigation flow.

Pumping stations: Pumping stations are facilities including pumps and equipment for pumping fluids from one place to another.

Weir: A small dam across a channel for the purpose of diverting flow, measuring volume of flow, or reducing erosion.

16 Agricultural drainage

READING FOR COMPREHENSION

The agricultural drainage series covers such topics as basic concepts; planning and design; surface intakes; economics; environmental impacts; wetlands; and legal issues.

WHAT IS AGRICULTURAL DRAINAGE?

Agricultural drainage is the use of surface ditches, subsurface permeable pipes, or both, to remove standing or excess water from poorly drained lands. During the late 1800s, European settlers in the Upper Midwest began making drainage ditches and channelizing (straightening and reshaping) streams to carry water from the wet areas of their farms to nearby streams and rivers. Later, farmers increased drainage by installing subsurface drainage pipes generally at a depth of three to six feet. Until the 1970s, most subsurface drainage pipes were made from short, cylindrical sections of concrete or clay called "tile." That is why terms like tile, tile drainage, and tiling are still used, even though most drainage pipe today is perforated polyethylene tubing. When installing a subsurface drainage system, pipes are either strategically placed in a field to remove water from isolated wet areas or installed in a pattern to drain an entire field. In some areas, surface inlets or intakes (risers extended from underground pipes to the surface) remove excess surface water from low spots in fields (Figures 16.1 and 16.2).

WHAT IS AGRICULTURAL DRAINAGE NEEDED?

Many soils in the Upper Midwest, as well as soils in other regions of the U.S. and the world, have poor natural internal drainage and would remain waterlogged for several days after excess rain without artificial drainage. This prolonged wetness prevents timely fieldwork and causes stress to growing crops because saturated soils do not provide sufficient aeration for crop root development. The roots of most crops grown in Minnesota cannot tolerate excessively wet conditions for more than a couple of days. Soil conditions that make drainage a necessity for some agricultural lands include those with slow soil water permeability or dense soil layers that restrict water movement, flat or depressional topography and, in some areas, high levels of salts at the soil surface. Large areas of Minnesota would not reliably produce crops if artificial drainage systems had not been installed (Figure 16.3).

Farmers must make a significant financial investment when installing an agricultural drainage system. They are willing to make this investment for two major reasons:

DOI: 10.1201/9781003293507-16

FIGURE 16.1   A tile (By Mohammad Albaji).

1. Agricultural drainage systems usually increase crop yields on poorly drained soils by providing a better environment for plants to grow, especially in wet years.
2. The systems generally help improve field conditions for timely tillage, planting, and harvesting.

These two factors have improved agricultural production on nearly one-fifth of U.S. soils. The most recent USDA comprehensive survey of drained lands showed that in 1985, 30% of all agricultural lands in the Upper Midwest (Illinois, Indiana, Iowa, Michigan, Minnesota, Missouri, Ohio, and Wisconsin) were artificially drained. Minnesota has large areas of poorly drained soils: e.g., 66% and 59% of the soils in the Red River and Minnesota River basins, respectively. In recent years, farmers have installed as much as 100 million feet of subsurface drainage pipe in Minnesota annually. A significant portion of new drainage activities is replacement and enhancement of old drainage systems. As old systems age and decay, replacement activities will likely continue.

FIGURE 16.2 Poorly drained agricultural land immediately following a storm; (http://www.extension.umn.edu/agriculture/water/publications/pdfs/issues__answers.pdf).

FIGURE 16.3 Subsurface drainage pipes are typically placed at depths of 3 to 4 feet in poorly drained soils, Ahvaz, Iran (By Mohammad Albaji).

Although agricultural drainage has benefited agricultural production in many regions and countries, there are concerns about its potential environmental impact. Subsurface drainage systems have a positive impact because they generally decrease the amount of surface runoff, thereby reducing the loss of substances generally transported by overland flow. There are concerns, however, about the potential negative impacts of drainage on the hydrology of watersheds, the water quality of receiving water bodies, and the amount and quality of nearby wetlands.

HYDROLOGY

Drainage systems are designed to alter field hydrology (water balance) by removing excess water from waterlogged soils. There are concerns about the downstream hydrological effects caused by draining this excess water. Anecdotal evidence indicates that streams and ditches have become "flashier" over time, spilling over their banks and causing localized crop damage. Some research articles suggest that the most dramatic hydrological changes in a landscape occur when it's converted from native vegetation to agricultural production, and that subsurface drainage may reduce peak flows in some situations. A recent regional publication summarized the environmental impacts of subsurface drainage on agricultural land. The authors concluded that subsurface drainage reduces surface runoff by 29% to 45%, reduces peak flows from watersheds by 15% to 30%, and has little impact on the total annual flow from watersheds. A publication that summarized drainage studies from several countries concluded that subsurface drainage generally decreases peak flows in fine-textured soils but often increases those flows in coarser, more permeable soils. This publication also found that subsurface drainage often increases base flow to streams. Locally based research is necessary, however, to better understand the impact that drainage can have at watershed scales. In addition, the impact of surface inlets on watershed hydrology is an important issue currently being examined.

WATER QUALITY

Surface drainage (enhancing overland runoff) tends to increase the loss of nutrients and sediment that occur with surface runoff. Subsurface drainage, however, can decrease surface runoff, thereby reducing sediment losses by 16% to 65% and phosphorus losses by up to 45%. The main water quality concern about subsurface drainage is the increased loss of nitrates and other soluble constituents that can move through soil to drainage systems and end up in nearby surface water. In addition, surface intakes, which are common across southern Minnesota and northern Iowa, provide a fairly direct pathway for sediment and other contaminants in surface runoff to reach nearby surface waters.

WETLANDS

Despite the fact that wetlands are protected by various regulations, it is estimated that over 60,000 wetland acres are lost nationally each year. The loss of wetland ecosystems – valued for their wildlife habitat, for water storage, and increasingly for their potential role to improve water quality is not easy to quantify. But it's likely that agricultural and urban drainage activities both cause wetland loss.

WAYS TO REDUCE THE POTENTIAL IMPACTS OF AGRICULTURAL DRAINAGE

Many current drainage research and Extension programs throughout the country are trying to find ways to reduce the potential environmental impacts of agricultural drainage while retaining its agronomic benefits. Some management practices have been effective; others are presently being examined. Both are described in the following sections.

IMPROVED NUTRIENT MANAGEMENT ON DRAINED SOILS

The proper management of crop nutrients (nutrient source, application rate, and timing) is an important way to help control the loss of nutrients through surface runoff and subsurface drainage water. It's been shown that the application of nitrogen fertilizer at rates higher than those recommended by the University of Minnesota increases the amount of nitrate removed through subsurface drainage systems. Since university recommendations are based on an optimum economic return, over application of nitrogen fertilizer should be less profitable. It should be noted, however, that drained agricultural soils have significant nitrate losses from the natural process of organic matter mineralization. Improved nutrient management can potentially reduce nitrate losses on drained lands by up to 30%.

CHANGES IN CROPPING SYSTEMS

Row crops such as corn and soybeans experience considerably more nitrate loss through subsurface drainage flow than perennials such as alfalfa and brome grass. So the incorporation of alfalfa or other perennials into farmers' crop rotations could significantly decrease nitrate losses to nearby surface water. While alfalfa may be a financially sound crop for some operations, it is not an economically viable solution for many Minnesota farmers.

IMPROVED DRAINAGE SYSTEM DESIGN

Subsurface drainage systems are designed to remove excess water from soil quickly enough to minimize crop stress in most years. Agricultural engineers have developed depth and spacing guidelines for installing drainage pipes. For example, recommendations for the many clay-loam soils prevalent in much of southern Minnesota call for placing drainage pipes approximately three feet deep and 60 feet apart or four feet deep and 80 feet apart. Either design should remove water at the same rate and give similar crop yields. It has been proposed that placing drainage pipes at shallower depths might result in less nitrate loss. This would happen because nitrate would be more likely to reach a biologically active but saturated zone and be converted to nitrogen gas by denitrifying bacteria. The conversion of nitrate/

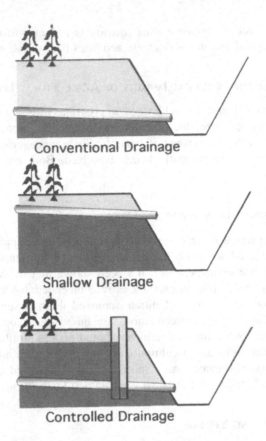

Conventional Drainage

Shallow Drainage

Controlled Drainage

FIGURE 16.4 Studies will determine if shallow drainage and controlled drainage reduce nitrate losses in Minnesota; (http://www.extension.umn.edu/agriculture/water/publications/pdfs/issues__answers.pd).

nitrogen to nitrogen gas would prevent the nitrate from reaching the drainage pipes and nearby surface waters. This practice, if proven effective, offers the advantage of being applicable anywhere that drainage systems are installed. It also requires no new management or capital investment (Figure 16.4).

CONTROLLED DRAINAGE

Controlled drainage has become recognized as an effective practice – and in other states, a best management practice – for mitigating nitrate losses from drainage systems. This practice involves placing simple water control structures at various locations in the system to raise the water elevation. This elevated water causes the water table in the soil to rise, which, in effect, decreases the drained depth of the field. Researchers from North Carolina, Ohio, Michigan, and Canada have demonstrated that controlled drainage decreases the volume of water drained (15–35%), slightly increases surface runoff (because soils have less space to store water), and significantly decreases (up to 50%) nitrate losses seen in conventionally drained fields. Decreases in nitrate losses have been attributed primarily to reductions in the volume of water drained and, to a somewhat lesser extent, by increased denitrification in the soil. If managed properly, controlled drainage has the potential to improve crop yields by making more water available to plants.

The application of controlled drainage techniques is limited, however, by topography. The process is economically unfeasible on land slopes greater than about one percent because more water control structures are needed as slopes increase. In addition, controlled drainage adds new management requirements to systems (also increasing with slope) that some will view as a disadvantage.

Surface Inlet Alternatives

Alternatives to the traditional "open inlet" are being used more frequently around Minnesota. One design involves digging a trench, placing drainage pipe at its bottom, and filling the trench with small rock. These "rock" or "blind" inlets slow the flow of water (compared to open inlets) and may reduce the amount of sediment reaching the drainage system. Another design involves the installation of subsurface drainage pipes in a very tight pattern in a small area in the middle of a wet spot. Another, more traditional, technique involves replacing open inlets with perforated risers. All these designs have the potential to do a better job of protecting water quality than open inlets, while still providing adequate drainage so crops don't "drown" (Figure 16.5).

FIGURE 16.5　A Drainage trencher, Ahvaz, Iran (By Mohammad Albaji).

WETLANDS

Wetlands have been proposed as a means of treating water from drainage systems before it is released into nearby rivers or lakes. Biological activity in wetlands can be effective at removing nitrate by converting it to nitrogen gas through a denitrification process that's similar to what occurs in soils. Researchers in Iowa suggest that wetlands can remove from 20% to 80% of the annual nitrate in subsurface drainage water depending on the ratio between the areas of drained land and wetland.

This approach to "treating" drainage water presents some challenges. Site topography may pose difficulties in getting subsurface drainage waters to the surface and into wetlands. Land requirements and the cost of construction are also important economic factors. Finally, the bulk of nitrate losses from drained lands in Minnesota occurs in early spring when wetlands are not functioning at their peak capacity to remove nitrate, because of low temperatures and high water flow rates. The potential effectiveness of wetlands in treating drainage water in colder climates requires more research.

EXERCISES

A. Read each statement and decide whether it is true or false. Write "T" for true and "F" for false statements.

TF1. A tile is a short, cylindrical section of concrete or clay.

TF2. The roots of most crops cannot tolerate excessively wet conditions for more than a day.

TF3. The most drainage pipes today are made from short, cylindrical sections of concrete or clay called "tile."

TF4. Drainage systems are designed to remove excess water from waterlogged soils.

TF5. Surface drainage decreases the loss of nutrients and sediment that occur with surface runoff.

B. Circle a, b, c, or d which best completes the following items.

1. Farmers increased drainage by installing .... at a depth of three to six feet.
   a. surface ditches
   b. subsurface drainage pipes
   c. controlled Drainage
   d. surface drainage pipes

2. .... involves placing simple water control structures at various locations in the system to raise the water elevation.
   a. Controlled drainage
   b. Shallow drainage
   c. Conventional drainage
   d. Subsurface drainage

3. .... have been proposed as a means of treating water from drainage systems before it is released into nearby surface water.
   a. Rivers
   b. Lakes
   c. Wetlands
   d. Sea

4. Subsurface drainage systems are designed to remove ....
   from soil quickly enough to minimize .... in most years.
   a. excess phosphorous – crop stress
   b. excess nitrate – crop stress
   c. excess water – crop stress
   d. excess salinity- crop water requirement

5. It has been proposed that placing drainage pipes at shallower depths might result in less .... loss.
   a. nitrate
   b. phosphate
   c. sulfate
   d. carbonate

C. **Match the sentence halves in Column I with their appropriate halves in Column II. Insert the letters a, b, c ... in the parentheses provided. There are more sentence halves in Column II than required.**

Column I	Column II
1. During the late 1800s	() **a.** can decrease surface runoff, thereby reducing sediment losses and phosphorus losses too.
2. Agricultural drainage systems	() **b.** is limited, however, by topography.
3. Subsurface drainage, however,	() **c.** is the decreased loss of calcium.
4. The main water quality concern about subsurface drainage	() **d.** usually increase crop yields on poorly drained soils.
5. The application of controlled drainage techniques	() **e.** is not dependent to the slope.
	() **f.** can increase sediment losses and phosphorus losses too.
	() **g.** European settlers in the Upper Midwest began making drainage ditches.
	() **h.** is the increased loss of nitrates.

D. **Cross out the word or words that make each statement false, and write the word or words that make each statement true in the blank.**

1. It has been proposed that placing drainage pipes at shallower depths might result in more nitrate loss.
   ....

2. Researchers have demonstrated that controlled drainage increases the volume of water drained (15–35%).
   ....

3. The controlled drainage is economically unfeasible on land slopes greater than about ten percent because more water control structures are needed as slopes increase.
   ....

4. Biological activity in wetlands can be effective at removing nitrogen by converting it to nitrate gas through a denitrification process that's similar to what occurs in soils.
   ....

E. **Give answers to the following questions.**
   1. What is agricultural drainage?
   2. What are the differences between conventional drainage and controlled drainage?
   3. Why are the wetlands important?

F. **For each word on the left, there are three meanings provided. Put a check mark (√) next to the choice which has the closest meaning to the word given.**

1. **Drainage**	irrigation	removes excess water	wetland
2. **Subsurface drainage pipes**	tile	collector	manhole
3. **wastewater**	fresh water	saline water	polluted water
4. **Waterlogged**	wetland	dry land	ponding
5. **surface ditch**	surface drain	surface pipe	surface tube

G. **Fill in the blanks with the appropriate words following from the following list.**
   water table   drainage   irrigation   rainfall   removal   farmers
   accumulation   floods

   The .... of excess water either from the ground surface or from the root zone is called .... Excess water may be caused by .... or by using too much irrigation water, but may also have other origins such as canal seepage or .... In very dry areas there is often .... of salts in the soil. Most crops do not

grow well on salty soil. Salts can be washed out by percolating
.... water through the root zone of the crops. To achieve
sufficient percolation .... will apply more water to the field
than the crops need. But the salty percolation water will cause the
.... to rise. Drainage to control the water table, therefore, also
serves to control the salinity of the soil.

H. **Read this passage and then circle a, b, c, or d which best completes the following items.**

There are two types of artificial drainage: surface drainage and
subsurface drainage. Surface drainage is the removal of excess water
from the surface of the land. This is normally accomplished by
shallow ditches, also called open drains. The shallow ditches discharge into larger and deeper collector drains. In order to facilitate
the flow of excess water toward the drains, the field is given an
artificial slope by means of land grading. Subsurface drainage is the
removal of water from the root zone. It is accomplished by deep
open drains or buried pipe drains.

The main purpose of this passage is ....

a. introduction of drainage.
b. introduction of surface drainage.
c. introduction of different types of drainage.
d. introduction of subsurface drainage.

I. **Translate the following passage into your mother language. Write your translation in the space provided.**

Subsurface drainage

Subsurface drainage is the removal of water from the root zone. It is
accomplished by deep open drains or buried pipe drains.

i. **Deep open drains**

The excess water from the root zone flows into the open drains (see
Figure 16.1). The disadvantage of this type of subsurface drainage is
that it makes the use of machinery difficult (Figure 16.6).

ii. **Pipe drains**

Pipe drains are buried pipes with openings through which the soil
water can enter. The pipes convey the water to a collector drain (see
Figure 16.2) (Figure 16.7).

Drain pipes are made of clay, concrete, or plastic. They are
usually placed in trenches by machines. In clay and concrete pipes
(usually 30 cm long and 5–10 cm in diameter) drainage water
enters the pipes through the joints (see Figure 16.7, top). Flexible
plastic drains are much longer (up to 200 m) and the water enters
through perforations distributed over the entire length of the pipe.

iii. **Deep open drains versus pipe drains**

Open drains use land that otherwise could be used for crops. They
restrict the use of machines. They also require a large number of
bridges and culverts for road crossings and access to the fields. Open
drains require frequent maintenance (weed control, repairs, etc.).

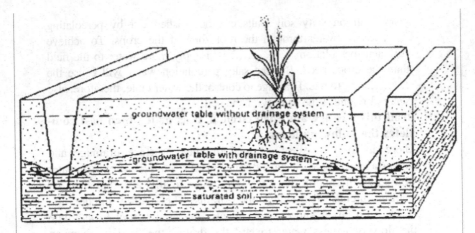

FIGURE 16.6 Control of the groundwater table by means of deep open drains; (http://www.fao.org/docrep/r4082e/r4082e07.htm).

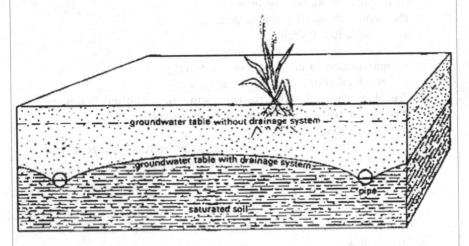

FIGURE 16.7 Control of the groundwater table by means of buried pipes; (http://www.fao.org/docrep/r4082e/r4082e07.htm).

In contrast to open drains, buried pipes cause no loss of cultivable land and maintenance requirements are very limited. The installation costs, however, of pipe drains may be higher due to the materials, the equipment, and the skilled manpower involved.

J. **Copy the technical terms and expressions used in this lesson. Then find your mother language equivalents of those terms and expressions and write them in the spaces provided.**

Technical term	Mother language equivalent
....	....
....	....
....	....
....	....
....	....
....	....
....	....
....	....
....	....
....	....

BIBLIOGRAPHY

Agricultural Drainage. Agronomy Monograph number 38:767-800. ASA, CSSA, and SSSA. (1996). Water Quality Guidelines. Agricultural Drainage. *Critical Reviews in Environmental Science and Technology* 24(1): 1–32.

Binstock, L. Minnesota Land Improvement Contractors. Personal communication.

Crumpton, W.G., & Baker, J. L. (1993). "Integrating Wetlands into Agricultural Drainage Systems: Predictions of Nitrate Loading and Loss in Wetlands Receiving Subsurface Drainage." In Proceedings: Integrated Resource Management and Landscape Modification for Environmental Protection, ASAE, Dec. 13–14, 1993, Chicago.

Glossary of Common Drainage Terms. Iowa Drainage Law Manual. pp 9.

http://www.directindustry.com/prod/mastenbroek-ltd/product-57769-1387863.html

http://www.extension.umn.edu/agriculture/water/publications/pdfs/issues__answers.pdf

http://www.fao.org/docrep/r4082e/r4082e07.htm

http://www.soils.umn.edu/research/.

Istok, J.D., & Kling, G.F. (1983). Effect of Subsurface Drainage on Runoff and Sediment Yield from an Agricultural Watershed in Western Oregon, U.S.A. *Journal of Hydrology* 65: 279–291.

Moore, I.D., & Larson, C.L. (1980). Hydrologic Impact of Draining Small Depressional Watersheds. *Journal of Irrigation Drainage* 106: 345–363.

Mulla, D.J. (1999). Minnesota River Basin Agricultural Resources and Research and the Red River of the North web pages at http://www.soils.umn.edu/research/

Randall, G.W., Huggins, D.R., Russelle, M.P., Fuchs, D.J., Nelson W.W., & Anderson, J.L. (1997). Nitrate Nitrogen in Surface Waters as Influenced by Climatic Conditions and Agricultural Practices. *Journal of Environmental Quality* 30: 337–344.

Randall, G.W., Huggins, D.R., Russelle, M.P., Fuchs, D.J., Nelson, W.W. and Anderson, J.L. (1997). Nitrate Losses through Subsurface Tile Drainage in Conservation Reserve Program, Alfalfa, and Row Crop Systems. *Journal of Environmental Quality* 26: 1240–1247.

Robinson, M., and Rycroft, D.W. (1999). In R.W. Skaggs and J. van Schifgaarde (eds.) "Agricultural Drainage." Agronomy Monograph number 38:767-800. ASA, CSSA, and SSSA, Madison, WI.

Skaggs, R.W., Breve, M.A. & Gilliam. J.W. (1994). Hydrologic and Water Quality Impacts of Agricultural Drainage. *Critical Reviews in Environmental Science and Technology* 24(1): 1–32.

Subsurface Drainage Studies in the Midwest. Ohio State Univ. Extension Bulletin 871.

Systems: Predictions of Nitrate Loading and Loss in Wetlands Receiving Subsurface Drainage." In Proceedings: Integrated Resource Management and Landscape Modification for Environmental Protection, ASAE, Dec. 13–14, 1993, Chicago.

U.S. Fish and Wildlife Service, "Status and Trends of Wetlands in the Conterminous United States, 1986 to 1997." Available at: http://wetlands.fws.gov/bha/SandT/SandTReport.html

Zucker, L.A., & Brown, L.C. (1998). Agricultural Drainage. Water Quality Impacts and Subsurface Drainage Studies in the Midwest. Ohio State Univ. Extension Bulletin 871.

GLOSSARY OF TERMS

Controlled drainage: Water control structures installed in the drainage outlet allow the water in the drainage outlet to be raised or lowered as needed. This water management practice has become known as controlled drainage.

Drain: A ditch and any watercourse or conduit, whether open, covered, or enclosed, natural or artificial, or partly natural and partly artificial, by which waters coming or falling upon a property are carried away.

Drainage: Four definitions may be used: 1) The process of removing surplus groundwater or surface waters by gravity or pumping; 2) The manner in which the waters from an area are removed; 3) The area from which waters are drained; 4) The flow of all liquids under the force of gravity.

Drainage system: A system of drains, drainage structures, levees, and pumping plants that drains land or protects it from overflow.

Subsurface drainage: Subsurface drainage comprised of drain tile or tubing designed to lower the water table by subsurface flow.

Tile drainage: The removal of surplus groundwater by means of buried pipes, with water entering through unsealed joints, perforations, or through surface inlets.

Watershed: The catchment area for rainfall that is delineated as the drainage area producing runoff. Generally considered as the area contained within a divide above a specified point on a stream.

Water table: The upper level of a zone of saturation in the earth, except where that surface is formed by an impermeable body (see perched groundwater).

Wetland: Generally, an area that has a predominance of hydric soils and is inundated or saturated by surface or groundwater at a frequency or for a duration that supports hydrophytic vegetation, typically adapted to those conditions. Wetland includes swamps, bogs, marshes, and similar areas.

Index

Printed in the United States
by Baker & Taylor Publisher Services